Iron and Steamship Archaeology

Success and Failure on the SS Xantho

Iron and Steamship Archaeology

Success and Failure on the SS Xantho

Michael McCarthy
Western Australian Maritime Museum
Fremantle, Australia

Kluwer Academic / Plenum Publishers
New York, Boston, Dordrecht, London, Moscow

Library of Congress Cataloging-in-Publication Data

McCarthy, Michael, 1947–
Iron and steamship archaeology: success and failure of the SS Xantho/Michael McCarthy.
p. cm. — (The Plenum series in underwater archaeology)
Includes bibliographical references and index.
ISBN 0-306-46365-2
1. Xantho (Steamship). 2. Shipwrecks—Australia—History. 3. Australia—Antiquities—Collection and preservation. 4. Underwater archaeology—Australia. I. Title. II. Series.

DU106 .M38 2000
623.2'204—dc21

00-035704

ISBN: 0-306-46365-2

233 Spring Street, New York, New York 10013

http:/www.wkap.nl/

10 9 8 7 6 5 4 3 2 1

A C.I.P. record for this book is available from the Library of Congress

Printed in the United States of America

Preface

In the early 1980s looting was occurring at the wreck of the iron screw steamship *Xantho* (1848-1872) and a request was made of the author to inspect the newly-found site and develop a strategy designed to put a stop to it. A test excavation was to be conducted and all loose attractive materials were to be removed and returned to the Western Australian Maritime Museum for conservation and safekeeping. It appeared to be a straightforward task—similar in ethos to many preemptive excavations conducted by the museum's Department of Maritime Archaeology in previous years.

In 1980, barely a year after the wreck had been found, the well-known British maritime archaeologist and theoretician Keith Muckelroy had stated that studies based on early steamships and the like, while interesting and sometimes providing useful display materials for museums, were not archaeology. He argued that "as an academic discipline" archaeology becomes redundant at the point where archives, representations, and oral histories provide more cultural information than can be obtained from the materials themselves. Further, he argued that the onset of "industrialisation and modern-style bureaucracies in the early 1800s marked the cut-off point" for underwater archaeological studies (1980:10). The influential maritime historian, David Lyon, then of the National Maritime Museum at Greenwich, was of a similar opinion. He believed that where detailed historical records of a particular ship were known to exist, there was little point in spending large amounts of time and money recording it's features on the seabed (cf. Henderson, 1988; Lyon, 1974). Certainly there was a great deal of literature on *Xantho* in the archives.

It had been built as a paddle steamer in Scotland in 1848 for use in sheltered waters, its original builders' contract and specifications were extant, and it appeared in a number of underwriters' registers. Over the course of its long career it received three registration certificates, including one of 1871 that recorded its conversion from paddle to screw propulsion. To add to this wealth of information, when it was lost on the Western Australian coast in November 1872, the incident was recorded at a court of inquiry in very great detail. As a result, it appeared at first glance that an examination of the wreck might not yield much new information about the vessel and the people involved.

On the other hand, another of the acknowledged leaders in the field, George Bass, had advised of "the value of archaeological research on ships recent enough for photographic records to be available" (1972:10). This divergence of opinion from two very well–known and highly respected maritime archaeologists was pertinent, for at the time the *Xantho* program commenced, the study of iron and steamship wrecks was arguably in its infancy.

The *Xantho* presented an opportunity to examine these divergent opinions when a brief scrutiny of the available historical records posed a number of interesting and unexpected questions. The most pressing was, what strange man would buy an ageing recycled paddle steamer in Scotland (especially one originally designed for use in sheltered waters) and then transport it halfway across the world to the sparsely populated and poorly serviced northwest coast of Australia? It was a remote place that was almost as close to Singapore as it was to any port of significance on the Australian coast. The owner/operator's decision-making powers were rendered the more questionable after the author's first inspection of the site in 1983. There it became evident that *Xantho* had been converted to screw propulsion utilizing an outmoded and inefficient, secondhand ex–Royal Navy horizontal trunk engine. Such coal-hungry machinery would not normally be expected on a vessel destined for the pearling grounds in the north of Western Australia—a remote place where fresh water for the ship's boiler was difficult to procure and where there were no coal supplies.

The first Europeans had settled in the region less than a decade before *Xantho* arrived and, as a result, there were no port facilities and the nearest marine workshops were in Surabaya and Batavia (now Jakarta).

Figure 1. A contemporary chart showing the proximity of the northwest Australian coast to Singapore and the major ports of Surabaya, Batavia, Makassar, and Kupang (derived from the Camden Harbour Pastoral Association Prospectus, 1864).

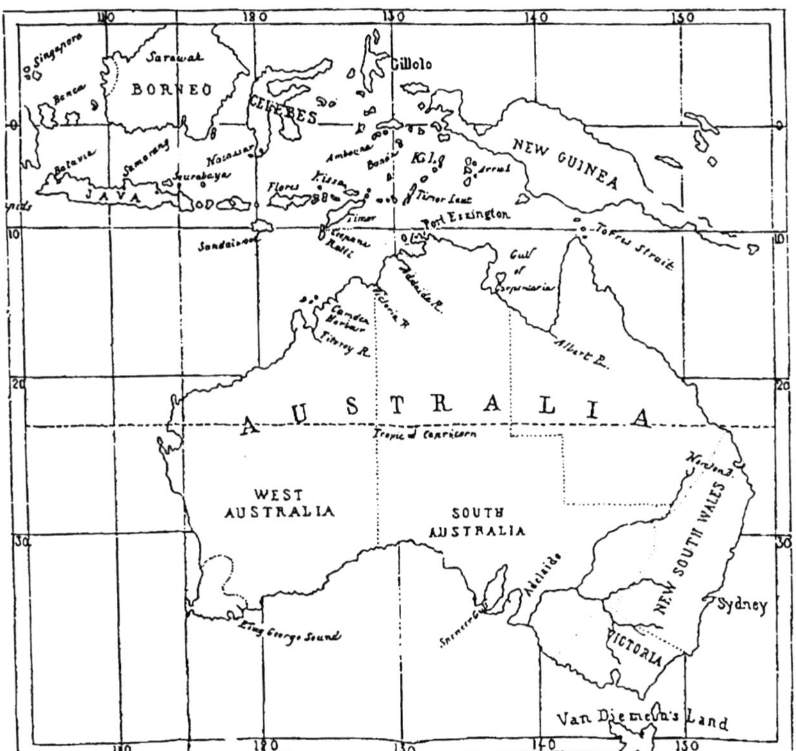

The available research also indicated that economists and historians alike had justifiably dismissed both *Xantho* (1848–1872) and its owner, the Englishman Charles Edward Broadhurst (1826–1905). *Xantho* sank beneath him only a year after he had purchased it for use in the Australian pearling industry, for example, and it soon became lost to living memory. Broadhurst himself was very controversial and was widely considered to be a persistent and somewhat unsavory failure.

The author's subsequent research (McCarthy, 1990) indicated that, though widely shunned by his peers, Broadhurst was in the mold of the widespread and very active nineteenth-century "British entrepreneur" as characterized by Payne (1974). He was also widely acknowledged for his propensity to go out of the "ordinary grooves" in search of wealth and was later described as a visionary possessed of all the decisiveness believed to be characteristic of the "typical American" (Kimberly, 1897). Because they also failed to understand its central role in Broadhurst's schemes, *Xantho* also appears to have been too readily dismissed by the historians.

These considerations together with the anomalies apparent in the material record ensured that much of the next decade was spent attempting to address them. This was achieved through a combination of archival research and the "excavation" of both the stern of *Xantho* and its odd engine.

The *Xantho* program also represented an opportunity to address other issues then becoming apparent with iron, steel (as a derivative of iron) and steamship wrecks, for in 1983 they appeared to represent a new class of underwater archaeological site. Thus the program was also partly driven by questions specifically aimed at investigating an iron-hulled steamship as distinct from the wooden-hulled vessels generally the object of research in maritime archaeology at the time. The questions included whether iron hulls generally behaved in the underwater environment like their wooden counterparts. Would they last as long? Were they as strong as they seemed? Did iron hulls pose any new problems to maritime archaeologists? Were traditional recording methods suitable for high-profile sites like iron and steel steamers, where hulls, boilers, and engines often dominate?

In combining all of these elements in a program now well over a decade old, the *Xantho* project evolved from its original site management and artefact-oriented brief into a much broader research program, with three recognizable phases: description, analysis, and explanation.

Acknowledgements

Key personnel on the *Xantho* project have been diver/rigger Geoff Kimpton, corrosion specialists Dr Neil North, Dr Ian MacLeod, conservator/restorer Dick Garcia, engineer/adviser Noel Miller, model-maker Bob Burgess, project artist Ian Warne, iron and steam shipwreck specialist John Riley, photographer/divers Pat Baker, Jon Carpenter, underwater video operator Lyall Mills, and film maker Ray Sutcliffe.

Dr Malcolm Tull of Murdoch University, launched me into the post-graduate study of Charles Broadhurst in 1985. The Masters thesis that he so thoroughly supervised was a corner stone to the historical and academic elements of the program to 1990. Associate Professor, Dr Peter Veth of James Cook University, supervised the Doctoral thesis on which this work is based and in doing so he also provided the academic and philosophical framework needed to help the *Xantho* project evolve from its intuitive base. Professor Richard Gould of Brown University, Dr Paul Johnston of the Smithsonian Institution and Professor John Campbell of James Cook University assessed the thesis, suggested it could be published, and provided many useful comments towards that end. Further editorial assistance was provided by Nick Burningham, Dena Garratt, Peter Harvey, Bill Jeffery, Vivienne Moran, Larry Murphy, Mike Nash, David Nutley, Dr Ian Oxley, Mark Staniforth, Shirley Strachan, Corioli Souter and Tom Vosmer. Sarah Stephenson and Bob Sheppard scrutinised the final draft from a 'public' context and made many important suggestions in preparation for its submitting to J. Barto Arnold,

Eliot Werner and support staff, notably Sandra Beris and Rosemary Sheridan at Kluwer/Plenum Publications.

Many more people have been involved in this project and many of their names appear in text, or alongside illustrations and diagrams. Those not mentioned there are field archaeologists, conservators and technical specialists Brad Duncan, Nic Clarke, Steve Cushnahan, Dena Garratt, Peter Harvey, Mike Lorimer, Brian Marfleet, Sally May, Colin Powell, Brunhilde Prince, Bob Richards, Fairlie Sawday, Mark Staniforth, Shirley Strachan, Jill Worsley, and Peter Worsley. Those members of the deconcretion team not appearing in text or in photographs are Carmela Corvaia, Jan Davies, Louella Doran, Brad Duncan, David Gilroy, Ian Godfrey, David Kelly, Alan Kendrick, Bob Richards, Vicki Richards, Fairlie Sawday, Rhonda Wozniak and Kent Jarman. Technical and professional advice and research assistance, again not credited in text, was provided by Lucy Burrow, Sue Cox, Chris Dixon, Lindsay Hill, Jeremy Green, Brian Kearns, Nikki King-Smith, Richard McKenna, Cyrus Mistry, James Pang, Ed Punchard Julia Redwood, Myra Stanbury, Richard Tomlin, Margaret Triffitt, Joanne Valley, Roland Webb and Laurie White.

General field and diving assistants at *Xantho* were J. Banfield, T. Beaver, P. Boardman, E. Boogard, B. Busby, M. Clarke, S. Cunningham, J. Dobbin, D Gilroy, J. Gloriod, J. Greenlees, D. Hardstaff, M. Hewitt, K. Hoffman, M. Jesser, B. Martindale, C. Martindale, S. McKenzie, D. McCarthy, M. McKay, D. McKenzie, R. MacLeod, J. Mercer, M. Moffett, P. O' Toole, J. Paskulich, J. Sewell, L. Simmons, J. Turner, G. Wallace and B. Williams.

Corporate Assistance was provided by ALCOA of Australia, Ansett Airways, Bill Busby (Budget Rentals), Canberra TV, Commonwealth Industrial Gases, Coates Hire, Main Roads Department, Perth Diving Academy, Prospero Productions, State Engineering Works, Society for Underwater Historical Research, Target Minerals, United Transport, Trans Australia Airlines and the Queensland Museum.

My family Debbie, Kim, Katie, Ellen and Phillip, provided all the support necessary to enable me to continue this work, both in the field and at home. The assistance rendered by Charles and Eliza Broadhurst's grand-daughter, Marjorie Darling and her extended family, especially Jane Brummitt, was also of fundamental importance to the project, as was the support of the Directors and Staff of the Western Australian Maritime Museum and the Western Australian Museum.

Contents

Chapter 1: Introduction

Chapter 2: *Xantho* and Broadhurst in context

Chapter 6: The investigation continues in the archives

Chapter 7: Excavations at the site

Chapter 8: Excavation in the laboratory

Chapter 1

Introduction

An evolving research base

1. DESCRIPTION, ANALYSIS AND EXPLANATION AT A STEAMSHIP WRECK

Because it had its roots as a site management program with a specific artefact and site-oriented brief, the *Xantho* project commenced in 1983 with a traditional artefact-oriented or *historical particularist* focus as defined by Jeremy Green (1990:235). It was an approach that was also strongly endorsed by George Bass (e.g., 1983) and it was considered "particularly appropriate" for the archaeology of shipwrecks, due to the fact that the materials were not yet well understood (Green, 1990:235).

1.1 Description

Despite its initial site and artefact management focus, the *Xantho* project represented an evolutionary step in modern maritime archaeology, for it was to become the first linking of biologists, corrosion specialists, and archaeologists underwater at a wreck *ab initio*. They were joined together by the author, not just for the usual artefact and site conservation purposes, but with a view to expanding in a scientific fashion onto what Gibbins (1992: 82–85) later referred to as Muckelroy's "most productive focus" on data characterization and site formation analyses.

Further, with respect to the divergent Muckelroy/Bass positions mentioned in the previous chapter, it was evident that the *Xantho* project could well be categorized as a form of historical and industrial archaeology. The latter is a branch of archaeology recognized and noted for its ability to validly utilize "eyewitness accounts...verbal descriptions...as well as historical and photographic records" (Renfrew and Bahn, 1991:271). Thus the use of historical documentation alongside the archaeological record (cf. Little, 1992) was a key element of the study.

In utilizing these sources, it was also understood from the outset that the historical record for people like the controversial Broadhurst was inevitably biased as a result of individual and familial forces and possibly due to "Victorian" and "post-Victorian" perceptions of respectability. An important consideration then was the activities that "structured" (developed, retained, destroyed, and/or modified) the documentary records.

With respect to the need to use archaeological evidence and written material as both complementary and potentially conflicting databases, it was also accepted that the archaeological and documentary records are both "imperfect representations of the same underlying reality" (Potter, 1992:10). As a result, the archival record was used as both an independent database and a source for generating alternative hypotheses about nautical behaviors and about Broadhurst (the individual), which could then be tested through the application of archaeological data. To a lesser extent, oral histories also provided a useful insight into Broadhurst and his activities, as will be seen.

In this manner the *Xantho* program attempted to utilize, yet move beyond, the relatively safe ground of particularist description, artefact analysis, and site management in maritime archaeology, to actively examine the people involved and to seek a valid explanation of their behaviors.

1.2 Analysis and explanation

It became evident at the 1983 site inspection that *Xantho* could be viewed as a poorly engineered vessel, unsuited for the inadequately serviced coast of Western Australia. As a result, its owner and operator Charles Broadhurst could be considered as a misguided and easily dismissed colonial entrepreneur—an impression that was supported by the available archival evidence, as indicated earlier.

Initially, the conclusions that the author reached about Broadhurst's behavior were intuitively based (e.g., McCarthy, 1985, 1986a). In later acknowledging this deficiency, there followed an attempt to make the processes of reasoning more explicit. It was noted that despite the "widely acknowledged" aim of processual archaeology—to seek valid explanation for the archaeological record—there was "very little agreement about explanation itself, about what constitutes a meaningful explanation, or about appropriate ways of validating or testing explanations that have been offered" (Renfrew, 1982:5). Richard Watson summarized the dilemma faced by those operating on anything but a descriptive and/or analytical level when he noted that "the harder it is to confirm or disconfirm hypotheses about a subject matter, the more interpretations can be given of it" (1990:73). These words attested to the perceived difficulty experienced by anthropologists in coming to a consensus about research strategies, validation, and processes of explanation (e.g., Bintliff, 1992; Courbin, 1988; Gould, 1990; Thomas and Tilley, 1992).

The use of excessive jargon was another problem in this period. Ian Hodder and others subsequently posed the question "How can alternative groups have access to a past that is locked up both intellectually and institutionally?" (Hodder, 1991:9). These self-evident concerns reflected the decades-old observation that overly "cabbalistic" language was a recognizable "disease of immature and growing disciplines" (Clarke, 1968:24). In the *Xantho* case, it was accepted nonetheless that to point to the transient and jargon-ridden nature of the theoretical approaches current in the 1980s and then dismiss them as passing fashions would have the "dampening effect on the growth of knowledge" identified earlier by Binford (1989:xiii). As Bintliff stated of contemporary archaeological debate, each age seemed to "write off" the achievements and understandings of our predecessors instead of proceeding in a "cumulative fashion, deepening our theoretical perspectives (1992:274).

In acknowledging these various issues, the *Xantho* research was conducted along the lines of Colin Renfrew's sentiments that, "the aims of explanation may be described, without initial reference to any explicit methodology, as to make intelligible" (1982:8).

Further, the period covered by the *Xantho* study saw not only the consolidation of processual archaeology in the mainstream and the emergence of the shipwreck anthropology movement in maritime archaeology (e.g., Gould, 1983), but also a strong defense of historical particularism. This occurred in maritime archaeology (e.g., Bass, 1983) and in some influential mainstream circles (e.g., Courbin, 1988). Later the *Xantho* study also saw the application of post processual approaches in maritime archaeology (as noted by Gibbins, 1992). In acknowledging these various movements in the late 1980s and early 1990s and in attempting to learn from them, the *Xantho* program evolved. In searching for a consensus, it was accepted that if relevant controls were applied to the examination of postdepositional effects (e.g., Schiffer, 1976; Muckelroy, 1978) some of the behavioral systems might be successfully identified. As a result, the requirement to control for postdepositional effects became a substantial part of the program.

An attempt was then made to further distil the evolving stages in accordance with Muckelroy's well-known and oft-presented model of the research stages in maritime archaeology (1978:249) and with those of other recognized theoreticians.

For example, "low level theory" or "empirical research with generalisations" (Trigger, 1989:19-24) includes the examination and analysis of the physical features and artefact assemblages at any archaeological site. These are the descriptive studies that provide the database—and much of the work conducted in the pre-1987 period at the *Xantho* may be categorized as such.

"Middle level theory" describes "generalizations that attempt to account for the regularities that occur between two or more sets of variables in multiple instances" (Trigger, 1989:20). The term has also come to encompass the study of site formation processes (e.g., Anuskiewicz, 1992; Gibbins, 1992). Herein lies the place of the study of commonalities in iron and steel ship disintegration and their application to the *Xantho* case (cf. Riley 1988a; Muckelroy, 1978).

Finally, Trigger's "high level theory" describes "abstract rules that explain the relationships among the theoretical propositions that are relevant for understanding major categories of phenomena" (1989:21).

At the seminal "shipwreck anthropology" seminar held in early 1981 (Gould, 1983), George Bass argued convincingly that social scientists rarely have the required command of "lower order data" on which to build "higher order inferences" in maritime archaeology, let alone to tap into competing theories of social order (Bass, 1983). The involvement of anthropologists in maritime archaeology during the 1990s ensured that some progress was made, however (e.g., Gould, 1995; Lenihan *et al.*, 1994; Souza, 1998).

Thus a re-working of the *Xantho* research questions occurred over time. This and the involvement of a growing group of specialists, including anthropologists, was consistent with two fundamental notions in archaeological theory. One, that "feed-back and constant reassessment" is required over the life of any archaeological project and two, that each of its stages can be "repeated and re-worked in the light of subsequent investigations" (Binford 1972:159; Muckelroy, 1978:249). In that regard, the longevity of the *Xantho* project has been a distinct asset. Some consider it one of the strengths of complex shipwreck excavations (e.g., Muckelroy, 1978:249).

In acknowledging that evolution occurred in the *Xantho* program throughout the 1980s and 1990s, it is useful to view the project within a general research design as articulated by Muckelroy (1978:249) before it commenced and then by Renfrew and Bahn as the fieldwork came to a close.

The latter authors describe four stages:

1. the formulation of a research strategy to resolve a particular question or idea;
2. the collecting and recording of evidence against which to test that idea, usually by the organization of a team of specialists and conducting of fieldwork;
3. the processing and analysis of that evidence and its interpretation in the light of the original idea to be tested;
4. the publication of the results in journal articles, books (Renfrew and Bahn, 1991:61).

Having posed the range of issues that surfaced as a result of the 1983 site inspection, a series of problem-oriented excavations were carried out on the *Xantho* engine in the laboratory during the period 1987–1995 (McCarthy, 1996). This was combined with a detailed archival study conducted in the period 1984–1990, focusing on Broadhurst (McCarthy, 1990). The recovery of both complementary and conflicting data from problem-oriented excavation and archival research served to encompass the first and second steps outlined above. The results of these inquiries were gradually synthesized and appraised against alternative reconstructions as each phase of excavation came to a close, to culminate in the third step. This is the equivalent of Muckelroy's stages of data analysis and "low-level interpretation". The exposure of the *Xantho* program to anthropologists in the early 1990s satisfied stage three above, or what Muckelroy described as "high level interpretation", or the "assessment of cultural implications" (1978:249). Publication, addressing stage four above, has been ongoing, commencing in 1985.

Finally, it must be noted that the inspection, excavation and management of the *Xantho* wreck have been the responsibility of a state museum. As a result the program was museum-based, causing it to have a strong public emphasis and to encompass the collection, research, education, public access, site management, and exhibition ethos of museum studies generally.

Chapter 2

Xantho and Broadhurst in context

Studies essential to an understanding of the material remains

To properly explain the significance of any archaeological site it is necessary to place it in its social, economic, and technological context. The site then also needs to be compared with similar sites in order to qualify and quantify any variance found. As Renfrew and Bahn note, "In order to reconstruct past human activity at a site, it is crucially important to understand the context of the find" (1991:42).

2. THE CONTEXT OF THE FIND

In the *Xantho* instance it is necessary to examine the state of marine engineering and iron shipbuilding in the period when it was built as a paddle steamer in 1848 and when it was refitted as a screw steamer and sold to Broadhurst 23 years later. These two analyses enable a picture to be formed of what is to be expected at the wreck of any iron ship of the period. Without the benefit of this step, it is impossible to make archaeologically valid comments on the significance of the remains and/or of any feature on the site that is considered anomalous.

In that context, shipping on the Western Australian coast needs also be examined in order to place the features found on the wreck into a regional, economic, and colonial framework and again to account for any variance found. An examination of Charles Broadhurst's entrepreneurial activities is then required in order to ascertain the place of *Xantho* and its contents within his business empire. In a later chapter, these understandings will be tested with the evidence taken from the wreck and further refined with evidence from the conservation laboratory.

2.1 Marine engineering

2.1.1 The iron hull

An iron hull can be expected on any wreck site formed in 1872 when *Xantho* was lost; for iron had been in use as a shipbuilding medium for over a half a century.[1] There are many histories available (e.g., Corlett, 1970; Grantham, 1859; Thearle, 1886). These show that the iron hull was seen to have numerous advantages to the ship owner of the time. These were strength combined with lightness (the latter producing less draught of water), greater capacity for stowage, safety, durability, economy of repairs, and less capital costs. The greatest disadvantage was seen to be fouling with weed and animal growth—a phenomenon that dramatically reduced the vessel's speed (cf. Corlett, 1970; Grantham, 1859). Builders, such as *Great Britain's* Brunel, knew little about the subject, however (Corlett, 1970). Insurers were equally disadvantaged. In 1843 when *Great Britain* was launched, for example, Lloyd's, the British Association of Underwriters, had just begun to collect information on iron ships with a view of insuring both them and their cargoes. By 1855 (seven years after the *Xantho* was built) and well after *Great Britain* had proved the worth of the iron hull, they had still not specified scantlings or mode of construction for iron ships. This reluctance was mainly because iron shipbuilding was considered to be still "in its infancy" and there were no "well-understood general rules" (Lloyd's Register, 1884:77).

The iron ship of this experimental period generally had a series of frames and deck beams in the form of single angle iron. This was, in effect, the application of European wooden shipbuilding tradition to the medium of iron. They were built, as a result, in the same fashion as wooden ships, with transverse frames and deck beams onto which the

[1] Iron and steamship archaeology includes wooden hulled steamers. Due to its focus on the SS *Xantho*, this work tends to concentrate on vessels with iron hulls.

planks were attached. The "longitudinal system" common to modern steel shipbuilding (e.g., Abell, 1948) was to come much later and thus lengthwise strength in early hulls was provided by decks, stringers, the keel, and the hull plates. Plates were relatively small and in many cases, including the *Xantho* and *Great Britain*, were applied in clinker (clencher or clincher) fashion. According to Thearle, the method was copied from the "well-known mode of fitting boat's planks, the lower edge of each strake overlapping the upper edge of the next lower strake" (1886:140). It needs be noted that Thearle (1886, Plate XI) shows the opposite arrangement, however. Further, on *Great Britain* the fifth strake from the keel is riveted flat onto the frames, with those above overlapping in clinker fashion as defined by Thearle and those below overlapping in opposite manner. See also Figure 39.

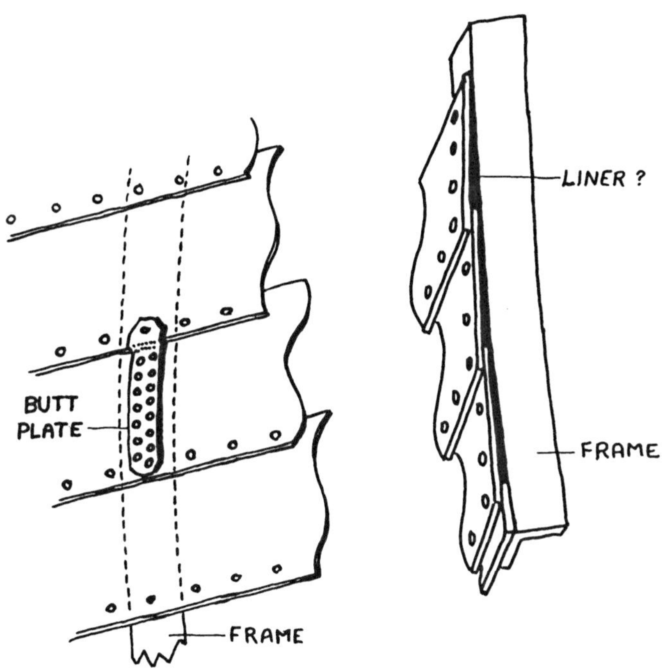

Figure 2. A sketch showing strakes of plating and tapered liners arranged as per Thearle's description. A frame is indicated by the dotted line. The external butt plate is similar to (XA 517) one later recovered from *Xantho* (Geoff Kimpton).

Between each plate and frame in the clinker system, there appeared a triangular gap, which required filling before the plates could be riveted to the frame. Sometimes "ill-fitting" washers were fitted around the rivet to close this, though a tapered "liner" or strip of plate was normally fitted between the frame and its adjacent plate. They were "difficult" to construct, and "commonly not very well made" (Robb, 1978:357). Thearle indicated that these problems, and the fact that the liners constituted yet another expense to the owners, ensured that the clinker system was superseded by what is variously termed the "in and out", or "raised and sunken" plating system. As a result, when Thearle wrote his treatise, it was rarely seen, except on "very old" iron ships (1886:140).

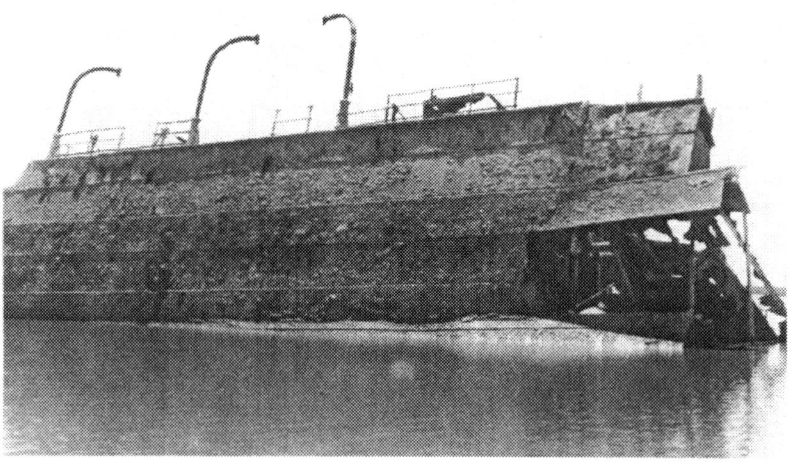

Figure 3. Hull plating amidships on SS *Colac*. The "in and out" system is clearly visible (M. McCarthy). See also Figures 19 and 39.

There was also considerable variation in the scantlings of vessels built in the experimental period of iron shipbuilding. Ocean going steamers were normally of heavy construction due to their size and operating parameters. SS *Great Britain*, for example, had relatively thick garboard plates of 11/16 of an inch (17 mm), with bilge and side plates of 10/16 of an inch (15 mm). *Star of India*, which was built in 1863, had even thicker plating (Wall, 1979; Reynard, 1979). Partly due to this over building (far in excess of the sizes later recommended) both of these ocean going vessels are extant and in fact the *Star of India* is still afloat. At the opposite end of the scale at the time, were small river steamers. The nineteenth-century iron shipbuilder John Grantham recommended that river steamers of the same size as *Xantho* had hull plating a maximum of 4/16–5/16 of an inch (c. 6–8 mm) thick for example (1859). Grantham and Lloyd's also set frame spacings at 18 inches (457 mm) for all steamers. Lloyd's also recommended watertight bulkheads 1/4 inch (6 mm) thick for vessels seeking a six-year sea going certificate. These are important statistics, as will be seen.

2.1.2 Marine propulsion

There are many treatises on the development of marine engineering, with Fincham (1851), Bourne (1858), and Burgh (1869) notable contemporary examples covering the period before *Xantho* was converted from paddle to screw propulsion. These and other sources such as Burgh (1873) and Jamieson (1897) also provide nineteenth-century analyses of the progression from paddle to screw propulsion and of the reasons for it.

Early steamers of *Xantho* vintage invariably carried masts and spars on which to carry sails, allowing their masters to utilize sail propulsion in favorable conditions. Drag of the paddles or the propeller then became a major concern. In these cases, an arrangement could be made whereby they could be feathered, or in the case of the screw disconnected and raised out of the water to counteract the resultant drag. On the other hand, a "dog clutch," enabling the propeller to be disconnected from the engine, and thereby freewheel in the vessel's wake, was often fitted if the screw was to be left immersed (see Figure 45).

Engines were occasionally fitted on sailing vessels for use as auxiliary propulsion where the wind was fickle or contrary, and there is provision in some underwriters' registers for this (e.g., *Key to the Register*, Lloyd's, 1869). Thus a nineteenth-century steamship could be found fitted with auxiliary sails, and sailing vessels, notably private "yachts" or other specialist craft, could be found with auxiliary engines. All can be found in the literature and in archives under the designation *steamer*, requiring some research into which mode of propulsion actually predominated (e.g., *Les Trois Amis* following).

In order to drive the new breed of screw-driven ships that began to steadily replace paddle steamers at sea in the mid-nineteenth-century, engineers originally attempted to adapt existing paddle engines to drive the screws. This caused considerable problems, for the screw required a greater speed of rotation (revolutions) than the paddle. To obtain the higher revolutions engineers employed a system of gearing to increase the speed of the slow-running paddle engines. The system was noisy, inefficient, and prone to failure. The following illustration shows the gearing on the engine of the SS *Royal Shepherd* (1853–1890) and the positioning of its vertical direct-acting, oscillating engine (a favorite among paddle steamer engineers in both its vertical and diagonal forms) immediately below the crankshaft.

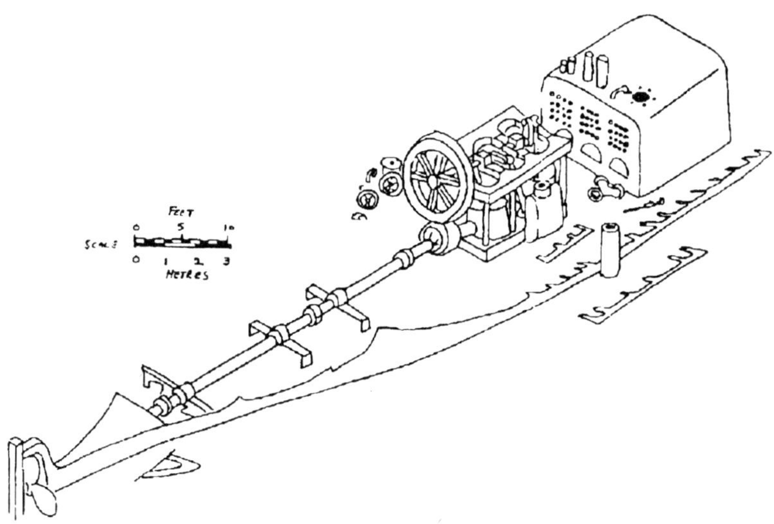

Figure 4. The stern section of the wreck of the SS *Royal Shepherd* showing its geared oscillating engine and a rectangular boiler (John Riley).

Engineers naturally moved away from these geared engines toward specifically built, high-revolution screw engines. As a result, jet condensing, horizontal engines, working to pressures of 20–25 pounds per square inch (p.s.i.) were fitted to the majority of naval and mercantile screw steamers in the decade after 1850 (see the discussion on engines and condensers following).

Though they were not remarkable for economy, these engines were much lighter and far more compact than the geared paddle engines that preceded them (Sennett and Oram, 1918).

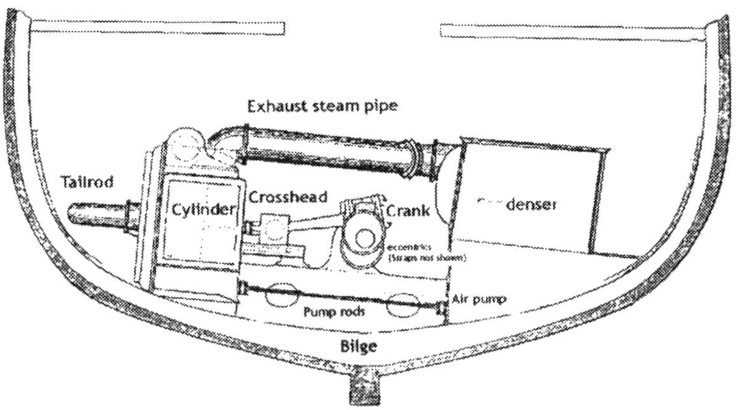

Figure 5. A horizontal engine derived from a similar design appearing in Sennett (1885). Note the crosshead or slide valve, the condenser and its pump rods (M. McCarthy).

Some of these marine engines came to be aligned in the vertical position, and being more easily accessible they became favored for their ease of maintenance. They were soon fitted to numerous ships, becoming known as the vertically inverted engine. This refers to the fact that though in the earlier vertical engines the cylinders were located below the crankshaft at the bottom of the ship (as shown in Figure 4), in later engines the cylinders were located above the shaft.

Although the vertical engine was seen to possess many practical advantages for merchant vessels, it was not introduced readily into the Royal Navy due to the necessity to keep machinery below the water line. Though there was interest in the vertical engine, in 1858 an admiralty committee decided in favor of horizontal engines because they could be kept low in the ship away from enemy fire. The horizontal types that the committee identified as being superior to the older types were the direct-acting engine, the double trunk (or two-cylinder trunk) engine patented by a consortium including John Penn (Patents for Inventions, 1855), and the return-connecting-rod engine made by Maudslay, Son and Field (Smith, 1937).

The trunk engine was designed to allow for the use of a relatively long connecting rod joined directly to the piston via a hollow trunk that projected through both ends of the cylinder. In eliminating the need for a crosshead, such as that shown above, the trunk engine had the capacity to become the most compact of all the direct-acting types. Indicating a personal preference among the horizontal type for the trunk engine, the engineer Burgh noted that Penn's trunk engine "combines simplicity of connection (piston to crank) and access for repair, with superlative design and arrangement" (1869:41).

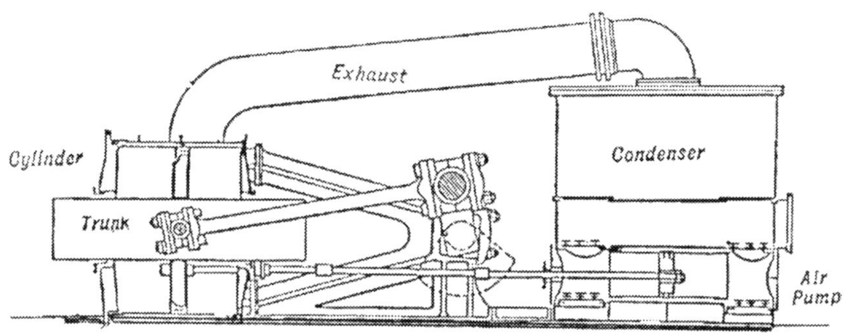

Figure 6. A trunk engine, showing the cylinder, trunk, condenser, and air pump. Note the long rods from the piston crossing the midline of the vessel to the condenser pumps (Jamieson, 1897:214).

The trunk engine type (which was built from 1846 to 1875) proved very expensive to operate, however. One contemporary engineer, Professor Andrew Jamieson, noted that the type became outmoded partly as a result of the exposure of its trunks to ash and other substances, to excessive friction at the trunk stuffing-boxes, and to heat losses on the exposed trunks (Jamieson, 1897). It was estimated that they used an average of 4–5 pounds (2–2.5 kg) of coal per indicated-horsepower-hour (Corlett, 1993:97), a prohibitive and uneconomical figure in the merchant marine. The SS *Himalaya*, for example, was built for mercantile use in 1853 with a 2500 horsepower (hp) trunk engine. It soon reverted to naval service as a troop ship having strategic (as opposed to economic) value by virtue of its size and mode of engineering. Excerpts of the engine-room log for 1855–56 are extant and these provide some idea of the problems faced by the ship's engineers (Guthrie, 1971).

Jamieson also noted that, because it had no intermediate slide valve or crosshead, the engine needed to rotate in a particular direction in order to minimize the combined effects of gravity, the angular thrust, and subsequent wear on the under surface of the piston. This is an important issue, as will be seen.

Thus the horizontal screw engine experienced its last popular usage within the framework and requirements of the Royal Navy and other major sea powers. It was still found in that context up until the late nineteenth-century, when it was rendered redundant partly due to the inefficiencies noted above and the development of the armored hull.

2.1.3 Boilers, condensers, and other machinery

Other developments were necessary before the screw engine was able to achieve its full potential. The first was the invention of a stern gland utilizing adjustable wooden inserts made from a very hard and self-lubricating timber called *lignum vitae.* Invented in 1854, this device solved the problem of keeping watertight the tube through which the rapidly rotating propeller shaft passed outside the hull to the screw (Barnaby, 1904). Before the device was perfected there were many problems with wear on the metal-to-metal surfaces. There are many illustrations of the stern gland (e.g., in Paasch (1890, Plate 25).

Secondly, the production of an efficient "thrust block" of the multi-collar type solved the problem of transferring the forward pushing force generated by the screw back along the propeller shaft. The thrust block, which consisted of a series of adjustable collars in an oil-filled container, effectively transferred the thrust from the rotating shaft via the enmeshing collars to a stationary block and then to strong bearers on the vessel's hull below (Figure 7).

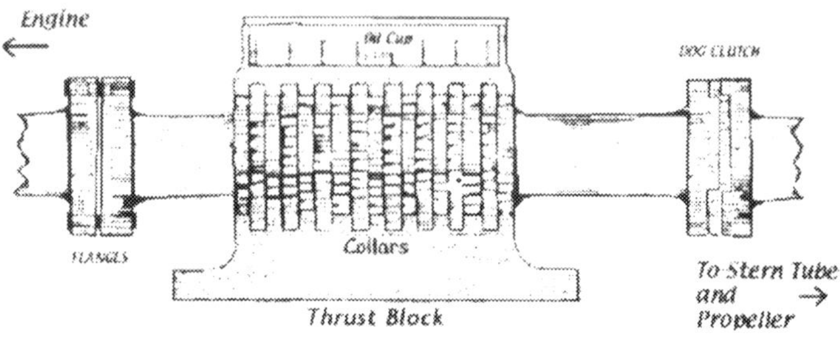

Figure 7. An illustration derived from a similar design in Paasch (1890) showing a section of propeller shaft, flanges, a multicollared thrust block, and a dog clutch (M. McCarthy).

Advances in boiler making were another prime consideration at this time. Though the production of steam for land engines and pumps dates back to the seventeenth-century, significant developments occurred in the evolution of boilers and boilermaking in the latter half of the working life of *Xantho*. These were well documented in the contemporary literature (cf. Burgh, 1873). Relevant to the *Xantho* is the fact that marine boilers initially used saltwater.

Though the steam emanating from saltwater was itself free of impurities, seawater requires more energy to be brought to boil. At one atmosphere, or 15 pounds per square inch (p.s.i.), a temperature of 213.2°F is required to boil seawater, as compared to 212°F for fresh water. As boiling continues and fresh water is taken off as steam to drive the engines, the remaining water in the boiler becomes progressively more saline. Unless an attempt was made to replace the water, its density increased and the required boiling point would also progressively rise, thereby requiring more coal. One simple method of achieving the best possible thermal efficiency was to note the temperature at which the water began to boil and if it was too high, to remedy the situation by replacing it with unused sea-water. It was recommended that if the boiling point of water reached a temperature of 215°F due to the increased salts, then it should be replaced with water from the ocean

(Main and Brown, 1855:28). Cold water as a boiler-feed was undesirable and thermally inefficient, however, a factor that could only be reduced if the water was preheated in a special vessel, called a feed-water heater.

Pressure was also a consideration, in that the higher the pressure in the boiler, the higher the boiling point of the water and the more salt was precipitated. At around 20 pounds per square inch pressure on the gauge (i.e., around 5 pounds per square inch above an atmosphere or 15 p.s.i.), the boiling point was relatively low and precipitation was minimal. As a result the use of seawater was not a real problem at low pressures, though there was a slow depositing of scale which required cleaning at regular intervals.

Being soluble over a wide range of operating parameters, sodium chloride (though it contributed to scale formation) was not as much a problem as was the sulphate of lime, or calcium sulphate, that was also in seawater. As boiler temperature rose, either through increased density of the saltwater or through the use of higher boiler pressures, the solubility of calcium sulphate decreased and it readily precipitated on the tubes, grates, and other internal surfaces. The precipitate produced a hard, poorly conducting scale obstructing heat transfer within the boiler itself, requiring more heat and therefore more coal to attain boiling point. In order to remove this encrustation, the boiler needed to be regularly shut down and cooled to enable the deposit to be physically removed from the interior.

Thus sulphate of lime in seawater was a major factor keeping boiler pressures down where saltwater feed was used. It was also well recognized in the mid-nineteenth-century that marine boilers using saltwater feed had a life expectancy of four or five years in comparison to a similar boiler fed with fresh water, which was expected to last 18 to 20 years (Bourne, 1858:83). A number of experiments designed to ascertain the cause of the problem and the value of using fresh water for the purposes of producing steam in the marine environment were conducted. Despite this, agreement was not reached on the causes of the corrosion by the time *Xantho* was refitted in 1871 (Burgh, 1873:356).

Another important device dependent on, and linked to, this problem was the condenser (see Figures 5 and 6). This facilitated the recycling of exhaust steam in the form of freshwater condensate by passing a jet of cold seawater over it (a jet condenser) or by passing the steam through, or over, tubes cooled by circulating seawater (a surface condenser). The jet condenser resulted in a mixing of the freshwater condensate with seawater and the surface condenser kept the seawater and freshwater separate. As a result, saltwater was not introduced into the system, with obvious thermal efficiencies at high boiler pressures. There were problems with surface condensers, however, in that the tallow and the other lubricants that entered the system through the steam chests and cylinders clogged the surfaces of the cooling tubes. These fats also decomposed at high temperatures into an acidic state, further reducing the life of the boiler.

Initially these problems limited the choice of condensers to the jet condenser. This sprayed seawater on the exhaust steam, cooling it so that the mixture of salt and fresh water from the recycled steam could be collected and reused. As coal consumption in a noncondensing engine of the time was calculated at 4 pounds weight of coal per indicated-horsepower-hour (Jarvis, 1993:156), a jet condenser was clearly better than having no condenser at all. The consumption rate of 4 pounds of coal per indicated-horsepower-hour, incidentally, is the same as that quoted earlier for a trunk engine fitted with a condenser, giving some indication of the relative inefficiencies of that type of machinery.

The rapid cooling of the exhaust steam in the condenser caused a near vacuum in the condenser and in the pipes leading to it from the engine cylinders. This was a phenomenon that was used to advantage in reducing the back pressure of the exhaust steam on the pistons by one atmosphere, or 15 p.s.i. Conversely, where steam exhausted straight to atmosphere, it immediately encountered a back pressure of 15 p.s.i., or one atmosphere. To remove it by use of a vacuum resulted in an increase in useable power and a saving in coal. On the negative side, pumps were needed to circulate the cooling water and to assist in maintaining or increasing the required vacuum (Yeo, 1894). Despite the need to power the pumps, an engine fitted with a condenser still proved far more thermally efficient than an equivalent engine without one. In recognition of this advance, the following comment was made in 1855:

> A noncondensing engine . . . will only be used where fuel is readily obtained and it is important to save space and weight. . . . [They] are serviceable for very short voyages in steamers . . . especially river navigation. . . . [The] condensing engine is more economical (Main and Brown:50, 67)

Thus a condenser of some sort was vital if long journeys were to be considered, if space was at a premium, and if coal supplies were limited. Jet condensers were fitted until the 1860s, due to the problems earlier mentioned with the deposition of fats and their breakdown into acidic compounds. The development of temperature-stable mineral oil lubricants in 1856 (Corlett, 1993) provided the breakthrough necessary for the further development of the surface condenser and the replacement of the jet condenser and saltwater feed. The transition proved quite rapid, so that by the end of the 1860s surface condensers were being regularly fitted to new or refitted steamers (e.g., *The Engineer*, June 17, 1898).

At sea, boiler pressures generally rose in unison with these advances—from 5 pounds per square inch in the 1830s, to 10 in the 1840s, 20 in the 1850s, and finally around 100 pounds per square inch in the 1880s (Smith, 1937).

The boilers that produced the steam themselves altered physically from the rectangular flue boilers of the 1840s to the rectangular multi-tubular boilers of the 1850s (see Figure 4) and then to the cylindrical multi-tubular type of the late 1860s. By the late 1860's, the cylindrical return-tube boiler, later to become known as the Scotch boiler, had proved so superior in regards to simplicity, reliability, and ease of maintenance that most other types became outmoded. The Scotch type then became the dominant form of boiler in the latter half of the nineteenth-century.

2.1.4 The advent of compounding

The achievement of much higher boiler pressures paralleled the development of efficient compound engines (Guthrie, 1971, chap. 6). These engines allowed high-pressure steam from the boiler to be expanded in stages; first in a high-pressure cylinder from where it was exhausted to lower-pressure cylinder(s) and from there to the surface condenser for recirculation back to the boiler. This allowed for substantial savings in fuel, reducing consumption to as low as 2 pounds per indicated-horsepower-hour (Jarvis, 1993:156; Griffiths, 1993:168–170). Less coal had to be carried, more space was available for cargo and fewer stokers were required. Compound engines powered by Scotch boilers not only used less coal and operated at a higher speed, but also weighed less and occupied less space. These advances and the opening of the Suez

Canal in 1869, through which the SS *Xantho* was soon to pass under Broadhurst's ownership, heralded the great age of the transoceanic steamship and the passing of the clipper ship era (MacGregor, 1973).

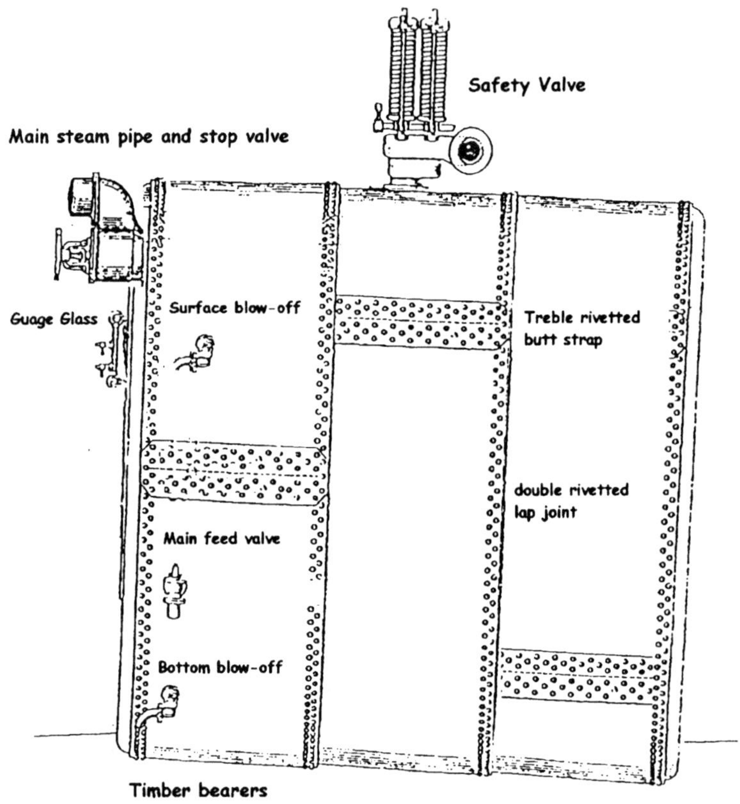

Figure 8. An illustration of a Scotch boiler with spring-operated safety valve. Derived from one appearing in Yeo (1894: 39). M. McCarthy.

With the Royal Navy convinced of the advantages of compound engines and of the value of the twin screw vessel, it became an obvious step for it to change from horizontal to vertical engines and to protect them with side armor. This allowed naval engineers the luxury, realized many years previously in the merchant marine, of tending to a vertical engine. Thus the replacement of the horizontal engine (whose only lasting saving grace was the ability to be kept below the waterline for tactical reasons) with vertically inverted compound engines was assured. The compound engine in its two-cylinder, triple, and quadruple expansion forms then took marine steam propulsion into the twentieth-century.

An indication of the rises in boiler pressures in British naval service is indicated in the table below. With the exception of the (clearly anomalous) high-pressure Crimean War gunboats of 1854–5, the link between the rise of pressure and the advent of compounding is evident. Again this is directly relevant to the *Xantho* case, as will be seen.

Table 1. The rise of boiler pressures in naval (RN) and merchant service (P&O fleet) against the advent of compounding. Distilled mainly from Corlett (1981:284), Tomlin, (1983), and Warsop and Tomlin (1990). note the position of the high pressure Crimean War gunboats (K. Kasi and M. McCarthy)

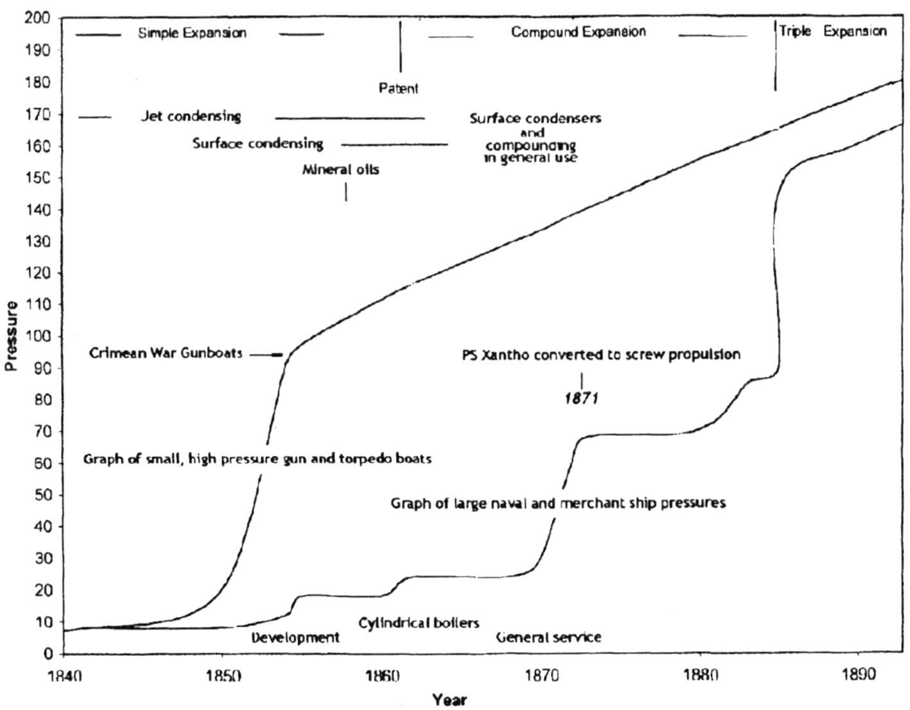

Figure 9 contains a dated selection of vessels utilizing Scotch boilers and (in most cases) compound engines. Those compound engines not buried or salvaged are identifiable by the different diameter of the covers on the cylinder heads (see also Figure 10). The position of the boilers, especially those that have rolled out of the vessel or lie on their ends is significant as will be seen.

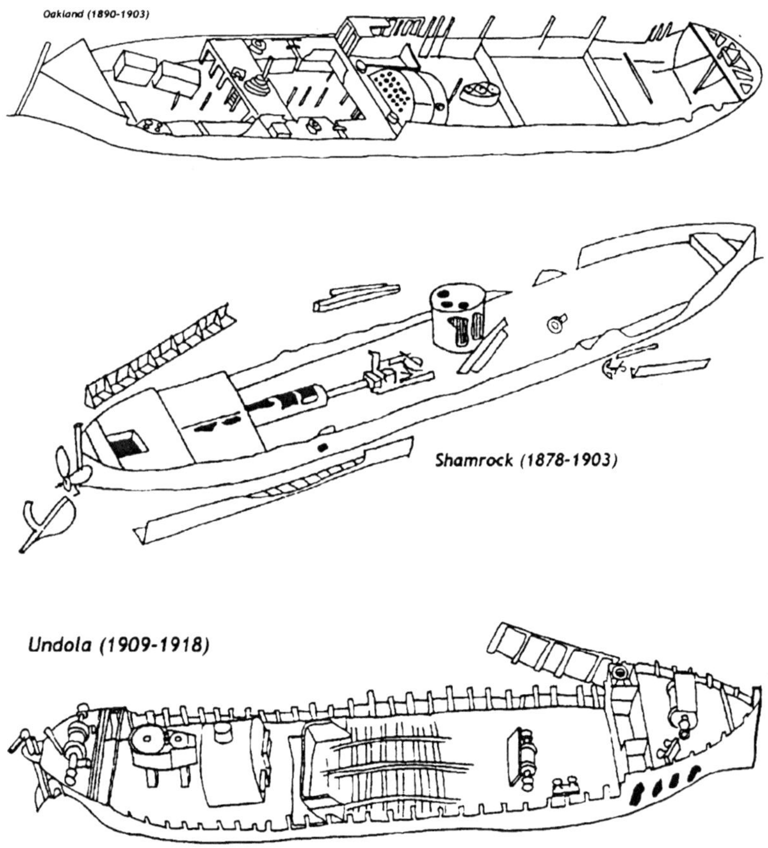

Figure 9. Sites with Scotch boilers and compound engines (John Riley).

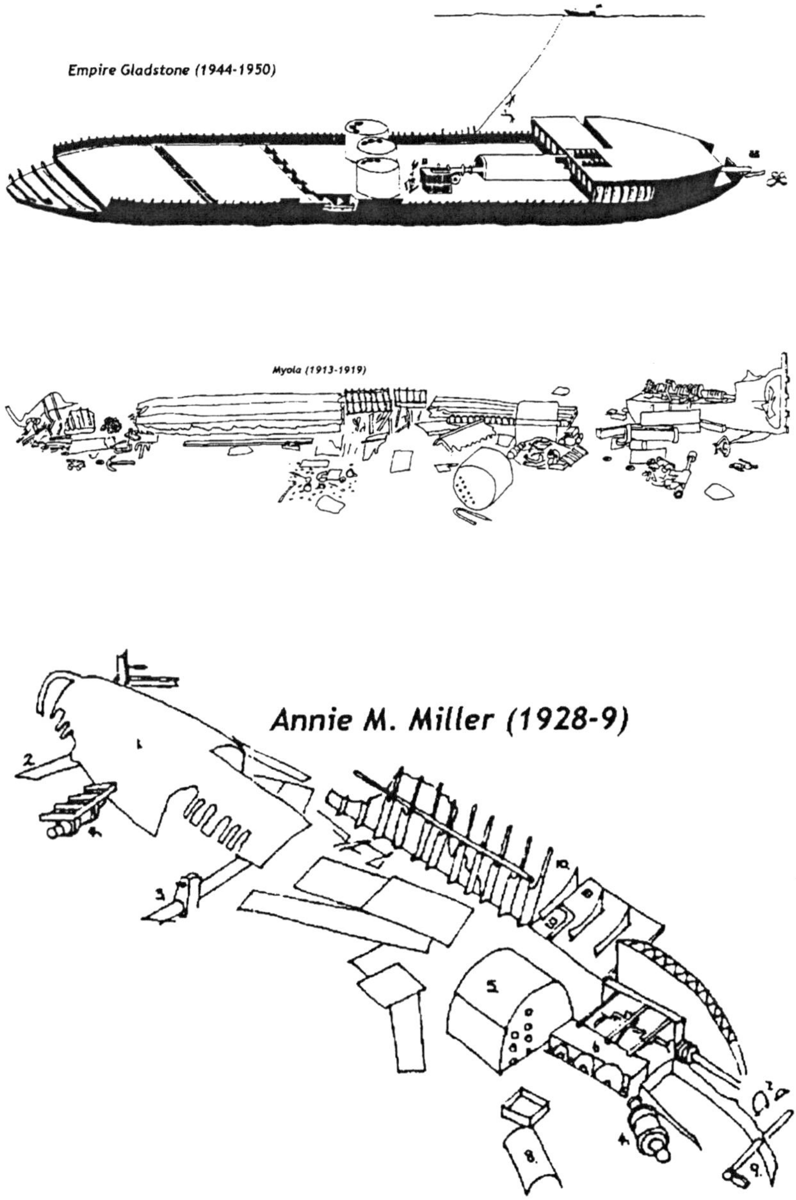

Figure 9 (cont.). Sites with Scotch boilers and (with the exception of *Woniora*) compound engines (John Riley).

2.1.5 The "ideal" 1870s steamer

In summary, the evidence presented above indicates that it is reasonable to expect that a newly engined or substantially refitted steamer of the 1870s that was intended for use over long distances on a coast with few facilities, or where coal supplies were at a premium, would display certain engineering features. These were an iron hull, auxiliary sails, possibly a lifting screw or alternatively a dog-clutch, a Scotch boiler, a surface condenser, a feed water heater, and a two-cylinder compound engine connected to a propeller via a multicollar thrust block and a stern tube lined with *lignum vitae.*

It needs be noted, however, that simple expansion engines were still to be found in effective service well into the latter half of the nineteenth-century. Normally they appeared only where coal was cheap, traveling distances were short, or if the capital needed to upgrade or purchase suitable "modern" machinery was lacking. These are found on the screw steamer (SS) *Royal Shepherd* (Figure 4, above) and the twin screw steamer (TSS) *John Penn* (1867–1889) shown in Figure 36, below, for example. These ships also carried rectangular boilers, as did the American blockade-runners PS *Nola* (1863–1864) and PS *Mary Celestia* (1864–1864) shown in Watts (1988). It also needs be acknowledged that paddles are found on steamers well into the twentieth century, as are wooden hulls.

Having addressed the developments in marine engineering that would have influenced the design and configuration of a ship like *Xantho*, built or substantially refitted in 1871, the regional factors that may have had an influence on its configuration and operation are now briefly examined.

2.2 The colonial setting

In his seminal work entitled *The Tyranny of Distance: How Distance Shaped Australian History,* the historian Geoffrey Blainey notes that Australia generally was slow to utilize the steamship even in its coastal waters and estuaries (1966). This, he believed, was due to the fact that the major cities on the coast were too far-spaced and the volume of traffic too small to warrant the cost in operating this type of vessel. In respect of steam on the west coast of Australia generally, another noted historian,

Franklin Broeze, believes that a critical shortage of capital was the most likely cause of the inordinate delay in the development of the Western Australian merchant fleet (1982). Another major drawback to the establishment of a coastal steam trade here was a chronic lack of coal. For these reasons the coastal trade in Western Australia was awaiting the advent of steam propulsion 20 years after steamships had captured over 30% of the traffic around Europe (Hartley, 1982) and nearly 50 years after Dutch-owned steamers entered service in what is now the Indonesian Archipelago (Figure 1). By 1829, when the Swan River Colony (Western Australia) was formed, there was a regular steamer link between the centers of Batavia (Jakarta) and Surabaya and overseas ports such as Singapore, for example (Roff, 1993). Further, experienced Western Australian colonial shipowners found themselves unwilling to enter the relatively expensive and complex business of steamship operations without governmental and logistical support in the form of a substantial subsidy. Indeed it has been claimed that they "may have shown sound business sense" in doing so (Hartley, 1982:97).

Thus, when the SS *Xantho* arrived at the colony in April 1872 to great acclaim as its first coastal steamer, it was almost half a century after the introduction of steamships in other parts of the world (Jamieson, 1897). In bringing *Xantho* to a colony without the infrastructure required of steamship operation, Broadhurst appears at a superficial glance to have made a great mistake. Further, with coal and engineering facilities in short supply or nonexistent, it would be expected that *Xantho*'s design would have been characteristic of sound, thermally efficient, and easily maintained steamers of the period.

2.2.1 Steamers in Western Australia

Michael Richards (1987) provides an examination of each of the steam vessels used in Australian waters, setting the scene for the analysis of these ships on a national scale. As Broeze (1998) indicates in his aptly titled *Island Nation,* however, the European settlements on the Australian coast were, in effect, distant "islands," often with social and economic characteristics of their own. Thus in order to ascertain what where the expected characteristics of a steamer, like *Xantho,* chosen specifically for the northwest, its contemporaries on that coast require scrutiny. There were none, for, as indicated, *Xantho* was the first steamer to operate on the northwest coast and as a result, attention needs be focussed on the types used on the Western Australian coast in general.

The 20 meter-long (65.7 feet) *Les Trois Amis,* an iron-hulled, 42-ton, clinker-built, schooner-rigged screw steamer arrived from Britain for use on the Swan River in 1855. It was newly built and had been fitted with a one-cylinder direct-acting 9 hp engine. Unsuitable as a river steamer, the ship was re rigged for operation on the coast as a schooner and it ran for a short time from Fremantle to Geraldton and Port Gregory, where it took on lead ore. Appearing in the press variously as "screw steamer" or more informatively as a "steam schooner" it used the engine only when necessary. In fact, the engine was used on only one known occasion after it left the river and that was to prevent the vessel going ashore. It was soon removed for economic reasons. In November 1858 *Les Trois Amis* was sold to one of Broadhurst's colleagues, the explorer and "blackbirder" (Pacific Island slaver) Francis Caddell and it serviced his nefarious interests until lost off Timor in 1884 (Dickson, 1993). It is interesting to note that by 1872 its iron hull was worn out and the vessel was clad in local timber.

The near-new 211-ton, 46-meter (151 feet) long, iron screw steamer *Georgette* was intended for use on a regular Government subsidized run on the southwest coast. Built in 1872 at Dumbarton for use as a collier, it was purchased for £14,000 and arrived in September 1873 fitted with a two-cylinder, condensing, 48 hp compound engine (Henderson and Henderson, 1988:211).

Georgette also had a large cargo-carrying capacity and two steam-operated winches. It appears to have been a wise choice, one that was engineered as expected for use on the Western Australian coast—with a two-cylinder compound engine, a Scotch boiler, spare propeller, sails, and a surface condenser. Despite this, the enterprise was dogged by misfortune and labor problems and *Georgette* was lost in 1876. The remains provide a useful physical comparison to *Xantho.*

It was replaced with the 163-foot (49.6 m) long, 267-ton, iron-hulled SS *Rob Roy.* Built in Scotland in 1867, it was originally powered with a two-cylinder, simple-expansion, low-pressure 50 hp engine. In 1872, it was lengthened and refitted with a two-cylinder, 60 hp compound engine, and in this mode *Rob Roy* proved a great success on the Western Australian coast.

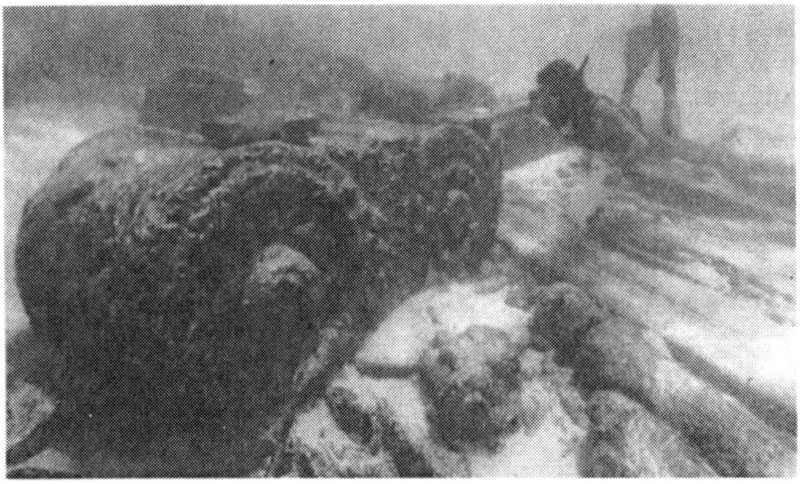

Figure 10. The SS *Georgette* compound engine, showing the differing diameters of the cylinders. The author is examining the high-pressure cylinder (Scott Sledge).

There are no others with which to broaden the sample of vessels examined against the *Xantho,* for in 1878 the shipping schedule was expanded to include an intercolonial service to Adelaide and Melbourne, thus requiring far more than small coasting or general-purpose steamers. These were the SS *Otway*, SS *Macedon,* and SS *Investigator.* Though their operating parameters were different, there is some benefit in examining their mode of operation, if only because the distances over which these steamers were operating were similar to those covered by *Xantho. Otway* was a 180-foot long (54 m), 271-ton iron vessel built in 1872 at Glasgow with a two-cylinder compound engine and *Macedon* was 220-foot long (67 m) and iron-hulled and was built in 1870 at Liverpool with a two cylinder 100 hp, simple expansion engine. SS *Investigator* was newly built, 210-foot (64m) long and fitted with a 97 hp compound engine and large cargo-carrying capacity. All, like the *Rob Roy* shown below, carried sails. One illustration is a copy of a contemporary painting (Page, 1975) and the other is the as it ship appeared on a dinner plate that was recovered during the author's excavation of a jetty used by *Rob Roy* during its colonial career (Garratt *et al*., 1995).

The variance in rig is of direct relevance to the debate on the complementary and conflictory nature of the historical (the painting) and archaeological (the plate) record. In this case the plate, which shows the ship rigged in full sail as a brig, appears to be an attempt to romanticize the vessel.

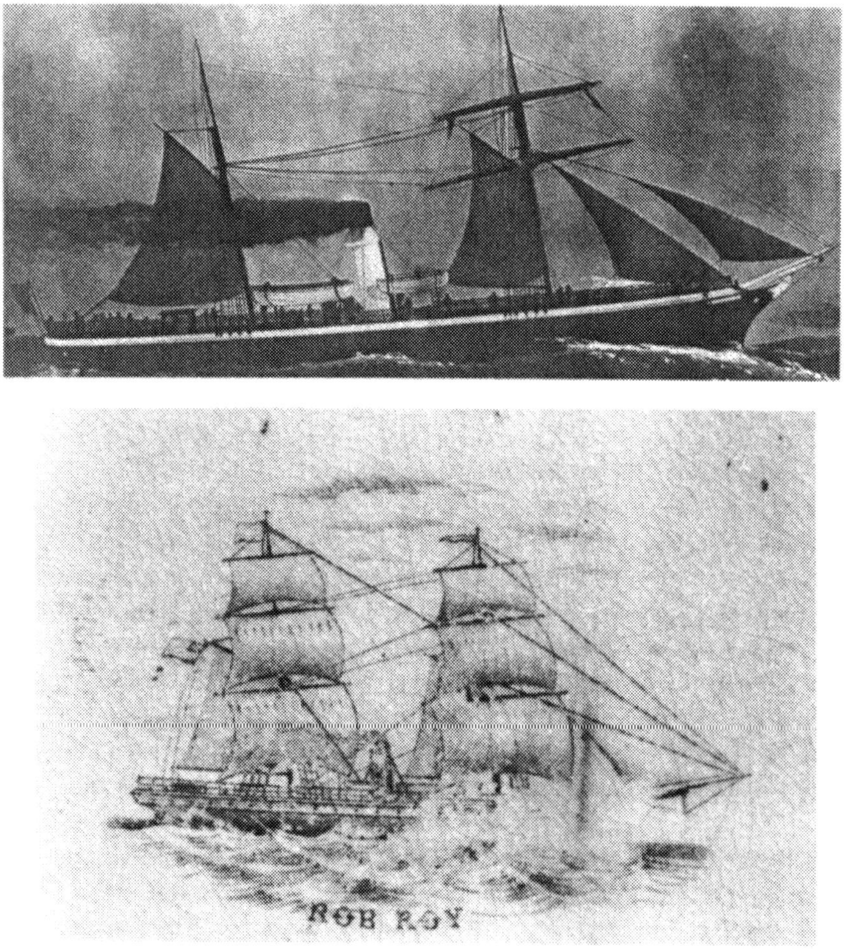

Figure 11a and b. Contemporary illustrations of the *Rob Roy* (Western Australian Maritime Museum).

Of importance in setting parameters against which to analyse the *Xantho* site was the fact that the ill-fated *Macedon* with its simple expansion engine was considered for a short-term charter only. Further, though consideration was given to replacing it with the SS *Claud Hamilton,* a vessel that had been built in Britain in 1862 with a 100 nominal horsepower (nhp) simple expansion engine, it had a "healthy appetite for coal" and was thereby rejected (Parsons, 1980:26). This ship was later refitted with a compound engine to make it more cost-efficient.

In examining these vessels as a sample against which to compare *Xantho*, the engineering expectations for a newly purchased (though not necessarily new) steamer intended for operation on the remote and vast west coast of Australia or across large distances on the coast have been established. These were an iron-hulled screw-driven vessel with sails, powered by a two-cylinder compound engine with a Scotch boiler and surface condenser. Though not mentioned in any of the instances above (mainly because those details were not included in registers or other descriptions), it would also have a *lignum vitae* stern gland and a device for disconnecting the engine when under sail.

In completing this section of contextual analyses it now remains to examine Charles Broadhurst's business activities in order to determine whether he exhibited a behavioral pattern that will prove useful as a pointer to the evidence found at the wreck of his ship.

2.3 The Broadhursts in context: An inadequate record

When the author began the search for archival material that would help in an understanding of the man largely responsible for the anomalous remains at the wreck of *Xantho*, little other than a number of dismissive accounts, short resumes, and family transcripts were available (Kimberly, 1897; Drake-Brockman, 1969; Weldon, c. 1960). The reasons for this paucity of sources are many. One is the tendency for people to destroy or suppress records about ventures in which they have failed, something Broadhurst did with remarkable regularity. Another was the tendency among post-Victorian authors and historians to write only about those characters and activities that could be recast and resurrected with an aura of success and respectability.

Families definitely followed this trend and relatives often destroyed materials such as diaries, letters and journals that, in their estimation, shed poor light on the individual or the family concerned. There are many famous instances and Broadhurst was certainly controversial, as will be seen.

A full and detailed description of the social and economic context in which Broadhurst operated may be found in a previous work entitled *Charles Edward Broadhurst, 1826–1905: A Remarkable Nineteenth-century Failure* (McCarthy, 1990). In beginning this précis of that work, the people involved, personal reminiscences, and the results of oral historical enquiries made over the past decade are presented in order to illustrate the fact that the subject matter is so recent, sometimes so well documented and often very familiar (cf. Purser, 1992). A series of coincidences involving the author are also presented for the same reason.

When I first met Marjorie Darling she was a remarkable 94-year-old lady who could still remember her grandparents, Charles and Eliza Broadhurst, and was very interested in talking about them. With her hand shaking gently on the arm of the worn settee, rattling the cup in its saucer, she told of what she knew of them, of their children, her mother and father, and of the many relatives. Some were in the East, one held a scrapbook, another had photographs, some had donated items to the museum. Strangely, none held any inherited "family" property. "There was Jenny, Gwen, David, Margaret, and many more," she said to me. I eventually contacted them, each leading to another family member and another insight into their extraordinary past. It was slow yet satisfying work, especially as the family found a common link to Charles and Eliza Broadhurst through the *Xantho* and their Aunt Marjorie, my informant. She was the daughter of Charles and Eliza's eldest son Florance and was the last surviving member of the family to have a direct link with Charles and Eliza Broadhurst. To Marjorie Darling, Charles was a virtual stranger. In her words he was "never home" being always away off in search of wealth, leaving her grand-mother Eliza to cope with her family and the problems Charles continuously caused. Charles was something of an enigma to them all. Eliza Broadhurst, on the other hand, was the family's heroine.

There was an odd stroke of amazing good fortune. This presented itself in a hospital room while staff were busy tidying up around me while I was convalescing after an accident. "That's my great grandfather's letter you've got there," one nursing sister exclaimed, pointing to Broadhurst's name on a letter that lay among research papers scattered at the foot of my bed. She was Penelope, a relative I was yet to meet, and she led me to important company documents, a will and even more contacts.

One comes to expect such co-incidences when conducting research in historical and industrial archaeology—for, as indicated earlier, they are fields where one is constantly reminded of the immediacy of the subject. The incident above, for example, was preceded nearly a decade before I joined the museum, or even heard of Broadhurst, by an entirely accidental visit to a dimly lit bed and breakfast establishment at Bournemouth in England. Cold and weary after hours on a wintry road, for me it was a welcome respite, and it was made the more so the next morning when I saw the Western Australian Aboriginal name *Karrakatta* on the hotel's sign. It was not until I began research into *Xantho* that I realised this was once Charles and Eliza Broadhurst's family home (see page 124).

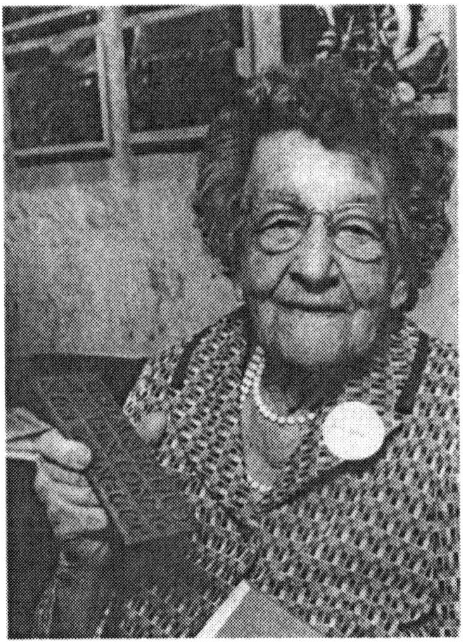

Figure 12. Marjorie Darling with the *Xantho* engine-maker's nameplate (*West Australian Newspaper*).

2.3.1 Victorian squatters

Charles Edward Broadhurst was born in 1826 in Manchester, England, into a privileged social and financial position that provided useful social contacts and the best possible education. All his brothers and sisters became well placed socially and financially. One, Frances Marris Broadhurst, married into wealthy circumstances in Glasgow, and another, Mary Louisa Broadhurst, married the famous and very wealthy engineer Sir Joseph Whitworth in the same year that the *Xantho* was purchased. This is doubly of significance, as will be seen.

In 1843 Broadhurst migrated to Victoria, in Australia, joining his elder brother on a vast pastoral holding at Kilmore, north of Melbourne (Hamilton, 1914). Again this was a privileged position and Broadhurst's emigration at 17 years of age into this wealthy pastoral context, surrounded by servants and laborers, would have further accustomed him to the life of a Victorian gentleman of considerable social standing and influence. Broadhurst was notably hardworking, however, and soon became a considerable success as a pastoralist and grazier in his own right. At 34 years of age and concerned by growing demands for land by newly displaced gold miners (Dingle, 1984), Broadhurst and his young Irish wife, Eliza Howes (1839-1899), turned their attention to newly-opened land in the north of Western Australia. It was an area that had just been described by explorers as a "pastoralist Eldorado" (Richardson, 1909:37).

2.3.2 Northwest pastoralists

Marjorie Darling could never really understand why the Broadhursts left Kilmore for the northwest. "Eliza was so happy in Victoria surrounded by her socially well-placed family and friends," she told me. Having endured hardship and adversity as one of 14 children born to a schoolteacher, Eliza developed a great love for the bustle of family life and the warmth of everyday society. She was practical and somewhat hardheaded and she loved teaching. She was also "a very talented musician with a good singing voice, " Marjorie Darling said.

It is interesting to note at this juncture that Eliza's beauty, and music and her social graces led to her family becoming an acknowledged social focus in the northwest. She and her husband also became close friends to the resident magistrate and his son (who was also his official assistant and personal secretary). Given that these two men were the sole representatives of government in the entire "North District" of Western Australia (Sholl diaries) this link was especially significant as it enhanced the standing and influence of the Broadhursts considerably. The district was a vast place encompassing the area north from the Murchison River (Figure 13). In 1862, just three years before the Broadhursts arrived, it contained not one known European inhabitant (McCarthy, 1990).

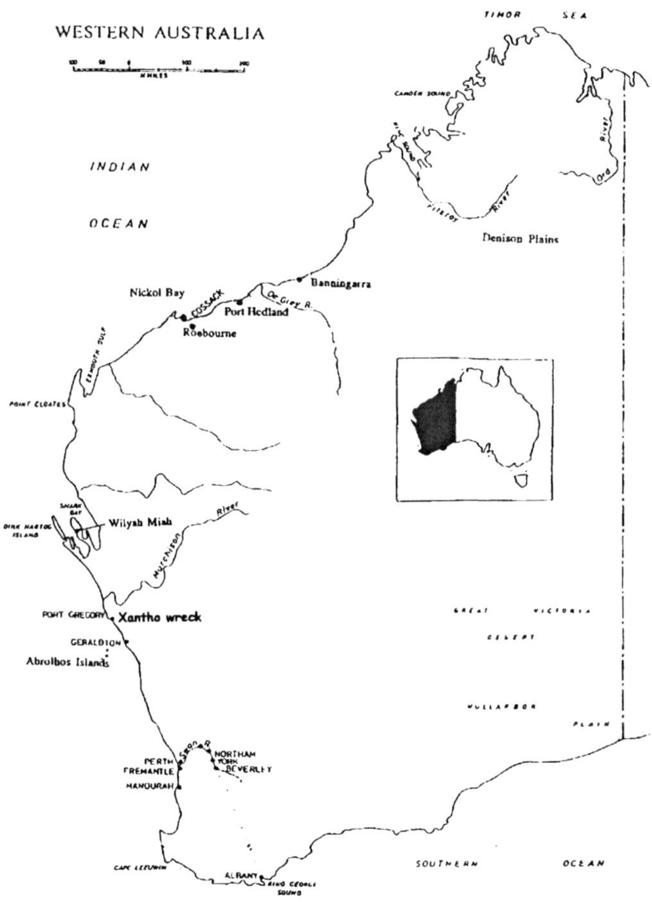

Figure 13. Western Australia in the mid nineteenth-century (Western Australian Maritime Museum)

As part of two speculative pastoral ventures, the Denison Plains Pastoral Company and Camden Harbour Pastoral Association, the Broadhursts were to land at Camden Harbour and proceed overland to the Denison Plains (Figure 13). There they were to establish a pastoral settlement with Charles as its leader in the field. This area, near present-day Halls Creek, had been much praised by earlier explorers unaware that they had traversed the country just after exceptional rains (cf. Grey, 1843; Gregory, 1884). The proponents of the scheme, one of whom was apparently Charles Broadhurst, envisioned that Camden Harbour and the Denison Plains were to be joined by a series of pastoral and telegraphic stations. These would then be pushed through to Sturt's route, which ran to embryo settlements on the north coast of Australia (near present-day Darwin) from Adelaide on the south coast. In this fashion overland links would be forged from Camden Harbour to the major eastern states capitals. On paper, the scheme (a brainchild of Melbourne-based land speculator William Harvey) was a sound notion which would shorten the route to Singapore, India and Britain and obviate the need for mariners to navigate the difficult Torres and Bass Straits to access Sydney and Melbourne (see Figure 1).

The journey from Camden Harbour to the Denison Plains was untried and dangerous, however, as letters published in a contemporary newspaper showed:

> Many, I fear will be induced to sell out their properties and pleasant homes to join Mr. Harvey's new Denison Plains scheme, when as surely as I write this letter many will die and leave their bones bleaching on the fiery heated rocks, long before their destination is reached (*Geelong Advertiser,* May 17, 1865).

The Denison Plains Company was somewhat of a mystery to Marjorie Darling before the author's study began. The family had told her and the other children very little about her grandfather Charles and his business ventures, as most degenerated into quite unsavory and very controversial affairs.

My delving did not worry Marjorie Darling, however, because a few details had filtered through to her as a small girl as she played around her parents" table and she had developed an interest in it. "Grandfather was somewhat of a dark horse," she would say, inferring that little was said about him due to his controversial past. Herein lies one of the major problems in relying solely on historical documentation biased by the subjective weeding out of what is considered socially acceptable by family and society generally.

This selective filtering of the documentary record is described as "organizational behavior" or the "conceptual category for the activities that have structured the ethnographic record" (Potter, 1992:10). It has obvious ramifications for the analysis of people like Broadhurst and needs to be constantly kept in mind. At no stage does the question of whether the trek from Camden Harbour to the Denison Plains was feasible appear to have been raised by the settlers before they departed Melbourne, for example. Broadhurst's decision to risk himself, the pregnant Eliza, and their two very young sons on the trek, attests to an apparent lack of common sense and the prevailing attitude of Victorian-era men to their wives and families.

Suffice it to say here that the settlement at Camden Harbour collapsed many months before the Broadhursts and their colleagues (including a doctor, surveyor, and servants) were due to leave Melbourne, but the delays in communication ensured that news had not yet filtered south. It arrived on the day they were due to depart and Charles Broadhurst was implicated in suppressing the news. They pressed on and he became a well-known figure within hours of first setting foot in Western Australia as the leader of what was undeniably the largest and best organised contingent of prospective settlers to pass through Fremantle on their way north. His influence was such that the government assigned to him an Aboriginal convict laborer and this began Broadhurst's involvement in another of the dominant and most controversial themes in his future life—his ill-treatment of laborers and staff.

They eventually landed at Nickol Bay (Figure 13), where there was a thriving European settlement, good land, and (initially) friendly Aborigines on whom to depend for labour and assistance. Mrs. Emma Withnell, the best known of the women settlers then in residence, passed down vivid memories of the company women's fine clothes which, in her words, were "similar to those illustrated in the magazines" (Withnell-Taylor, 1987:73). Her thoughts on seeing the seven months pregnant Eliza Broadhurst landing in her finery with two young boys and her piano in tow are unfortunately not recorded. That the Broadhursts transported the piano to the north, and apparently intended hauling it across the north of Australia on an untried route, is a statement in itself. Equally, as

Marjorie Darling indicated, the strong-willed Eliza would have refused to go without it!

Despite Broadhurst's efforts at holding it together, the company eventually collapsed, supply ships did not come, and a terrible drought set in. Despite their pressing needs, Broadhurst stoutly resisted pressure to break up the flocks and to feed the sheep to both his family and his people, creating great dissension among the group. This understandably engendered considerable ill feeling toward him and news of the confrontation filtered south when the first members of the failed company landed there. To many in the tiny settlements of Perth and Fremantle, Broadhurst was the person responsible for the debacle, and the collapse of the Company was to be the beginning of a long-lasting suspicion of him:

> Mr. Broadhurst is certainly a smart man . . . but if success be the test of ability he has certainly not proved himself superior having made a pretty mess of the Denison Plains Company . . . There is certainly a great distrust of him here (Barker and Gull letters, State Archive, 2423a, May 16, 1868).

A savage drought then brought further hardship to all in the north. This and the disappearance of a coastal schooner with nearly one third of the entire European population of the North District on board, caused Eliza and the children to leave at the end of 1867, never to return. After seeing them safely out of the northwest and on to a brief respite with their family in Melbourne, Charles returned to battle on at Nickol Bay, a shunned man.

It is believed that the dramatic failure of the Denison Plains venture and its effect on their social standing was perhaps the driving force behind Charles and Eliza Broadhurst's subsequent Western Australian colonial career and their ensuing 30-year search for regained wealth and social position. One could argue that it was a major influence in Broadhurst's decision to purchase the *Xantho* and to operate it in conditions so foreign to an English-born gentleman pastoralist.

2.3.3 Pearlers in the northwest

Pearl shell, or mother of pearl, fetched very high prices per ton, sometimes as much as £100 landed at London (Bain, 1982). This was equivalent to a mid level government servant's annual wage at the time (see Appendix 2). Pearling was both a lucrative and a brutal industry, however. Even at the time it was acknowledged by the press that suffering and slavery were the lot of the Aboriginal men and women who were "recruited" to provide the labour force for the European pearlers:

> The thirst for shells, for pearls for success, brutalises . . . the pearling speculator or diver. No day is respected, no dark man's life is valued . . . but the utmost of diving must be sucked out of them, killing them or not (*Inquirer*, April 28, 1875).

In expectation of great financial returns, all the northwest pastoralists engaged in what is known as *dry shelling* (harvesting pearl shell at low water spring tides) using local Aboriginal people as collectors (1998d).

As indicated, Broadhurst took his family south at the end of 1867 and in doing so missed out on what was to be a crucial formative period. Those who stayed learned more about the location of shell beds and the most efficient (though not necessarily the most humane) means of utilizing Aboriginal men and women in the industry. More importantly, the Aborigines were learning at a rapid rate how to make the transition from dry shelling or wading, into *naked diving* (i.e., diving without any aids such as goggles or fins). Naked diving began just as Broadhurst left for Melbourne and the transition was in full swing by the middle of the following year. By August 1868 divers were descending to depths of 6 fathoms, or around 10 meters (McCarthy, 1994).

Being excellent swimmers, with extraordinary eyesight even underwater, Aboriginal divers also used the currents to advantage by allowing themselves to be carried along over vast areas of seabed. Though they were without protective clothing and swimming aids and though they suffered terribly, the availability of cheap and plentiful labor made this a most productive exercise—one that is still practiced today by Indonesian fishermen in the trochus shell industry as shown in the following illustration (McCarthy, 1991).

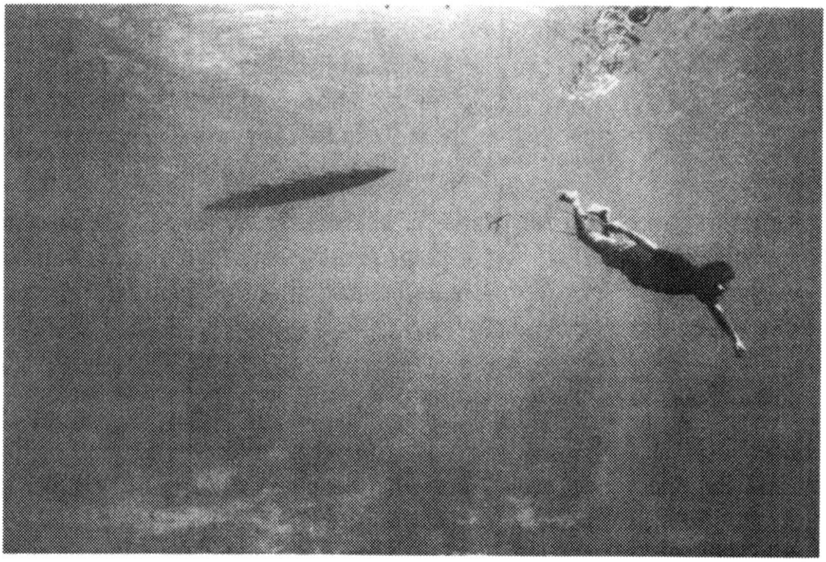

Figure 14. An Indonesian diver in 1991 using almost identical methods as those used by Aboriginal divers in the 1860s (Jon Carpenter).

2.3.3.1 Introduction of the "hard hat"

Broadhurst returned to Western Australia in April 1868 with plans to introduce diving apparatus or the *hard hat*, as it was also known, to the pearling industry. Though the use of this equipment was commonplace outside of Australia and the technology was well-known (cf. Davis, 1955) there were regional considerations to be taken into account before it could be successfully utilized.

Through sheer inexperience, Broadhurst and his colleagues chose a 27-meter long, wooden two-masted schooner that was far too large and then compounded the problem by diving in the narrow Flying Foam Passage at Nickol Bay. Though a rich source of shell, the passage was subject to very strong currents, sometimes in excess of 3-4 knots. They had great difficulty maneuvring the boat, and even with heavy lead boots, their diver was streamed out helplessly in the current. After a short while they gave up, having failed dismally in the attempt. The Aboriginal divers, in the meantime, used these same currents to great advantage, as indicated earlier. Broadhurst and his colleagues had attempted to apply unnecessarily complex technology, with inexperienced operators, on a ship that proved far too large for the local conditions. They were doomed to fail from the outset. More importantly, in the context of any attempt to understand the *Xantho* site, this and the Denison Plains venture provide some clues to Broadhurst's propensity to embark on schemes that "on paper" were feasible, but in reality were untried and speculative.

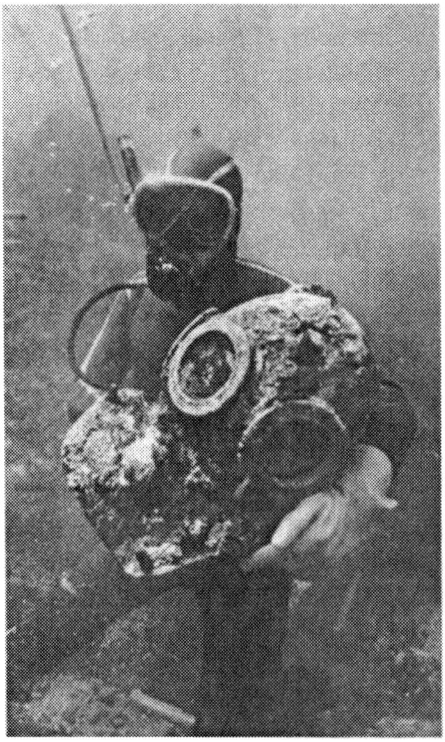

Figure 15. A "hard hat" found on the wreck of the SS *Macedon* (1870–1883). It was lost in a contemporary salvage attempt (Jeremy Green).

Despite the failure of this first known attempt to use diving apparatus in the Australian pearl fishery, Broadhurst correctly realized that the equipment had potential. He was 20 years ahead of his time in introducing it to the region, however, and there was a great deal to be learned to adapt methods to the tides and waters of the northwest (McCarthy, 1994).

Broadhurst clearly had vision in realizing the value of technological advances, but he did not understand that a great deal of experimentation was required before the technology could be efficiently applied.

2.3.3.2 Aboriginal convicts as pearl divers

By this stage pearlers were finding it difficult to acquire Aboriginal labor due to the increasing number of operators, the brutality of many, and the ravages of smallpox. Characteristically, Broadhurst looked elsewhere for labor and applied to interview Aboriginal prisoners at Rottnest Island, near Fremantle, with a view to enticing them to volunteer for service in the pearl fishery. There they were to work alongside the local Aborigines and his "hard hat" divers.

Despite some strong public opposition to the idea of Aboriginal convicts being released into the hands of the already notorious pearlers, Broadhurst was successful in obtaining 18 "volunteers" for the 1870/71 season. For reasons unknown, the Rottnest convicts proved hopeless in the sea and understandably tried to escape at every opportunity. Further, just as they found good pearling beds, the weather turned foul, becoming a series of gales that continued throughout the season.

Broadhurst traveled far and wide—to places identified through plant specimens collected on the voyage (T. Willings to McCarthy, October 9, 1991), but again he proved unsuccessful. Despite this, he felt that the newly found pearl beds were extensive and rich, requiring only the application of "labour and capital" to be efficiently harvested (Broadhurst to Sholl, April 14, 1871).

2.4 Broadhurst and *Xantho*

In mid-1871, Charles Broadhurst traveled to Britain with intentions of revolutionizing the nascent pearling industry in Western Australia by introducing capital, another new labor source, and advanced technology in the form of steam. When he left, 31 sailing vessels, ranging from one to 56 tons, were in operation on the pearling grounds along with 52 smaller boats or dinghies. All were wooden-hulled and not one was a steamship. Further, the sum of £4,500 that Broadhurst eventually expended in the purchase and fitting out of the *Xantho* was at least 30 times that of a mid to upper-level government servant's annual salary. It was a sum that would be measured in the millions of dollars today. (See a schedule of contemporary wages and salaries in Appendix 2.)

Where Broadhurst got the money is somewhat of a mystery. Pearling for him in the seasons before he purchased the ship appears to have been one series of disasters after another. The sale of his pastoral interests may have part-financed the purchase of the vessel, as could have the sale of his stock and land in Victoria. He may also have been a beneficiary in his recently deceased and wealthy sister Frances's will. She died in Glasgow in October 1871, the same month Broadhurst purchased the ship there. Equally, his family in England may have assisted in the purchase and even in the choice of vessel. When in England, for example, he would also have visited his newly married sister, Mary Louisa, who had just wed the famous engineer Sir Joseph Whitworth (Lee, 1900). For reasons that will later be made apparent, Whitworth may have advised his new brother-in-law that Penn and Son were excellent engine builders and that Glasgow, a noted shipbuilding region, was the good place to get a secondhand steamer.

Though he was to have only one steamer, Broadhurst intended to diversify his activities by using it as a "mother boat", servicing sailing dinghies that he had working at distant bases in the pearling industry on the northwest coast. He was also carrying five boats on the vessel's deck, together with 120 tons of coal, 30 tons of stores, whaling gear, and everything that his experience suggested was useful in order to pursue as many and diverse a range of maritime pursuits as possible. This included whaling and turtle shell collecting.

Apart from being harvested, the pearl shell also had to be transported to a suitable location for sale and manufacture. Broadhurst's intention to use a steamship to operate against time and tide in difficult pearling harbors and to satisfy the dual aims of harvesting and transporting shell and laborers between northwest Australia, Batavia, and Surabaya was sound. Port Hedland, where he established one of his pearling bases, for example, was considered a beautiful harbor, completely landlocked but not suitable for sailing vessels in anything but perfect conditions. With the steamer, Broadhurst could collect and deliver his shell with regularity. Finally, when not required for pearling, he intended to use *Xantho* as a *tramp steamer* or general-purpose carrier that does not run on any regular line or to a fixed schedule, but takes cargo "wherever shippers desire" (de Kerchove, 1948:853). Thus the relatively high costs of obtaining coal, employing qualified officers, and the many other expenses incurred by operating the SS *Xantho* could be recouped. It was a maximizing strategy, common and identifiable through much of human endeavour in a difficult environment or an untried economic context.

In utilizing illustrations of its contemporaries, registers, and other relevant details (including descriptions in the local press and the evidence given at the court of inquiry when *Xantho* was lost), the author has developed an impression of *Xantho* as purchased by Broadhurst. This conceptualization is presented by project artist Ian Warne in the following illustration. Also used in developing this impression were the results of the 1983 site inspection. These showed that the engine and boiler were located as far aft as possible (see Chapter 4). They also indicated that the boiler projected above the deck, requiring insulation in the form of square tiles. These tiles are visible in both site plans (Figures 32 and 45) and are conceptualized in Figure 25.

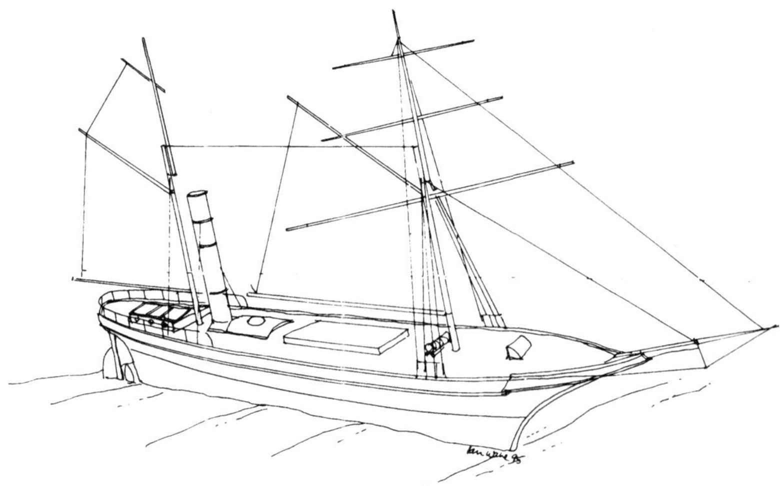

Figure 16. An impression of the SS *Xantho* in 1872 (Ian Warne).

2.4.1 Xantho: A transport for pearl divers

En route to Australia in *Xantho*, Broadhurst called in to his agents in Singapore, Batavia, and Surabaya. There he obtained coal, engaged 40 divers and continued onto his chief pearling base at Banningarra east of Port Hedland, arriving around April 1872 (Figure 13). It was a safe, virtually landlocked, harbor with two and a half fathoms of water (5 m) at low tide. Near a small freshwater lagoon not far from shore, Broadhurst established a camp with a substantial wooden house and "sick-bay" for his "Malay" divers. Firewood was also in plentiful supply.

The importation of Malays[2] as indentured labor began around 1871. It was a logical progression given that the supply of Aboriginal labor was drying up and the Malays had been operating as divers for centuries in the islands to Australia's north. Broadhurst transported over 140 Malays on *Xantho* alone and by 1875 nearly 1000 imported divers were recorded on the northwest coast (McCarthy, 1990:209–215).

In stark contrast to the Aboriginal people, who were mercilessly exploited, the Malays required to be housed, fed, paid, and then repatriated at the end of their employment. In an attempt to provide his men with a reasonable diet, and no doubt reduce his costs and profit from the supply of the food, Broadhurst successfully applied for land and planned to establish a 300-tree coconut plantation. As with most of Broadhurst's projects, his idea was innovative and the plan was good, but things started to go terribly wrong. One of his divers drowned and some of his men stole one of his boats from another base at Port Hedland and set off for home, for example. Others stole a boat from one of his bases at the Flying Foam Passage and also departed for home. Broadhurst's problems led to official concern and the police subsequently made an inspection of Broadhurst's pearling camps and found that:

> the Malays have not the slightest idea of either swimming or diving, being completely out of their element in the water. . . . I [was] . . . not the least surprised at their getting drowned (Votes & Proceedings of the Legislative Council, 1874–5, 714/57:168).

Broadhurst failed while others around him succeeded even though they were fishing the same grounds and using the same methods and the same labor. One author infers that Broadhurst was continually duped by all his laborers in the pearling industry (Bain, 1982).

[2] *Malays*: a term generally but incorrectly applied in the nineteenth-century to the people from places in present-day Indonesia, the Philippines, Malaysia and Singapore.

Some of his Malays were eventually employed as crew of *Xantho* and others (most likely the poorer divers) were employed as general hands or servants (see below).

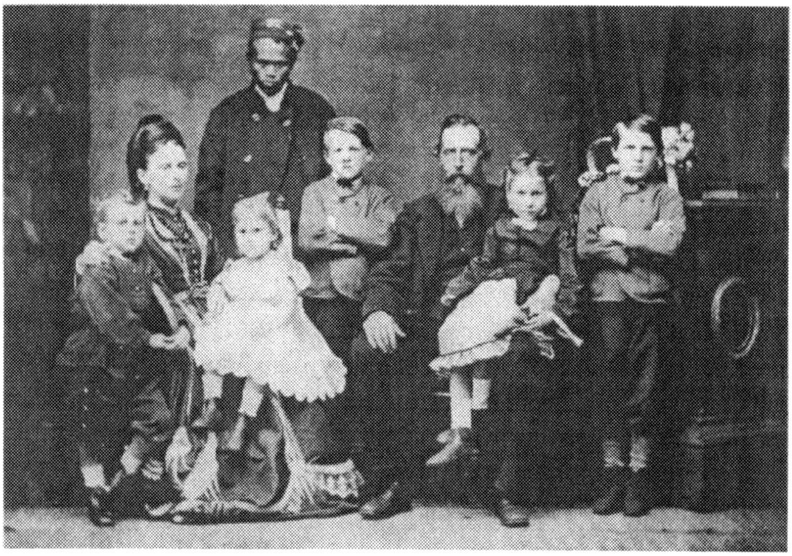

Figure 17. A family photograph of the Broadhursts and their Malay servant in mid-1872. Taken during one of Charles's infrequent visits home. They are in mourning dress due to the death of their baby Ernest (Broadhurst Family).

2.4.2 Xantho the tramp steamer: An indigenous record?

Xantho eventually completed two round trips between Fremantle, Batavia, Geraldton and Broadhurst's northwest pearling camps, carrying shell, goods, and Malay divers. When conditions allowed, Broadhurst also employed *Xantho* carrying passengers and assorted cargo and it transported a number of northwest Aboriginal men from the port of Cossack and the courts of nearby Roebourne to the Aboriginal prison at Rottnest Island near Fremantle.

As the first steamship to operate on the northwest coast, *Xantho* not only made a great impression on the newly-arrived Europeans, but also on indigenous groups like the Jaburrara, Martuthunira, Ngarluma and Yindjibarndi people (Horton, 1994) who had been living in the hinterland of Nickol Bay for many thousands of years. One group produced a number of rock engravings of a ship with smoke belching from its funnel.

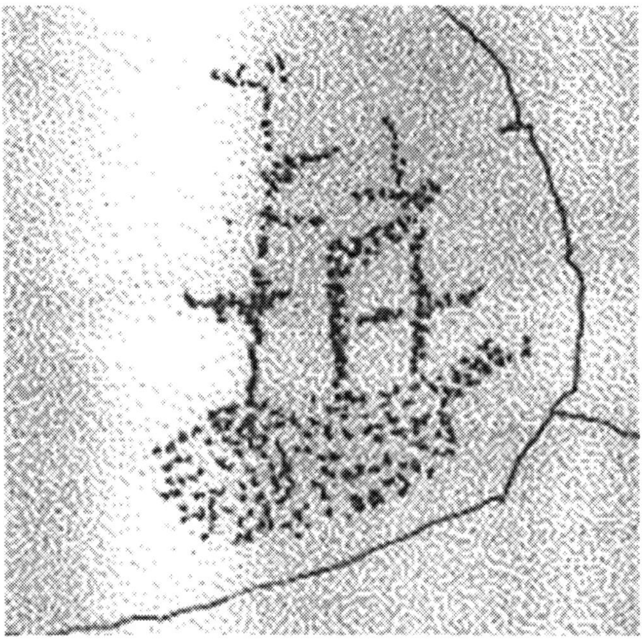

Figure 18. Ian Warne's impression of one of a series of nineteenth-century rock engraving showing a steamship believed to be *Xantho.* (The location of the engravings is not reproduced here in accordance with the wishes of the custodians of the site.)

The engraving appears in association with a number of other elements (e.g., a horse and cart, stock, a woman in a long skirt, and a man with a gun) of what was rapidly becoming a disruptive European presence. Though it could have been HMS *Cossack*, a ship that called in to the area around 1875 and gave its name to the chief pearling port in Nickol Bay, the steamer has been identified by one prehistorian as *Xantho,* a vessel with a profound visual and social impact on the Aboriginal people (Reynolds,1987). The men transported south, for example, appear to have been very senior "law men," obligated to respond to European transgressions of both aboriginal custom and their time-honored custodianship of the land (Noel Nannup personal communication, November 1999). They did so in a manner unacceptable to European law and most never returned home, however. (See Horton (1994) for a brief eluciation of the complexity of Aboriginal society and the customs and belief of Aboriginal people).

2.4.3 The last voyage

On its last trip the little steamer traveled from Batavia to Broadhurst's pearling bases, first Banningarra, then Port Hedland, and finally the Flying Foam Passage at Nickol Bay. At the Flying Foam Passage the ship was allowed to rest on the bottom at low tide and like all the other vessels in the vicinity it would have remained in that position until the next high tide. The tides in the vicinity are often in excess of 4 meters (Royal Australian Navy, *Australian Tide Tables*, 1999) and this often resulted in vast areas of the seabed drying at low water.

It was customary, while a vessel lay at rest on the sand or mud at low tide, to perform all the cargo and passenger handling functions using carts or bullock drays. Sometimes Aboriginal people and Malay divers were used as the workforce. This would have allowed them the opportunity to examine a vessel at close quarters and the total exposure of the hull would also explain why some of their illustrations of European craft show the vessel's underbody. (See for example Figure, 23.)

The practice of allowing even very large sailing vessels and steamers to lie "high and dry" whilst loading and off-loading was common on the Australian northwest coast until a just few decades ago. One well-known example is shown below (Simmer, n.d.).

When the tide rose at the Flying Foam Passage, *Xantho* was floated off and they departed for Geraldton as planned. While there, Broadhurst heard of a good sideline and immediately took the steamer back north to nearby Port Gregory to load a cargo of lead ore, which was lying stacked on the beach in sacks in readiness for transport to London. It was a narrow harbor with very strong currents and was notoriously dangerous for large overseas sailing vessels (Totty, 1986). It was, however, admirably suited for use by a small steamer such as the SS *Xantho* and once the cargo was off-loaded, the intention was to take *Xantho* on to Fremantle and to continue as a tramp steamer until it was needed back at the pearling grounds.

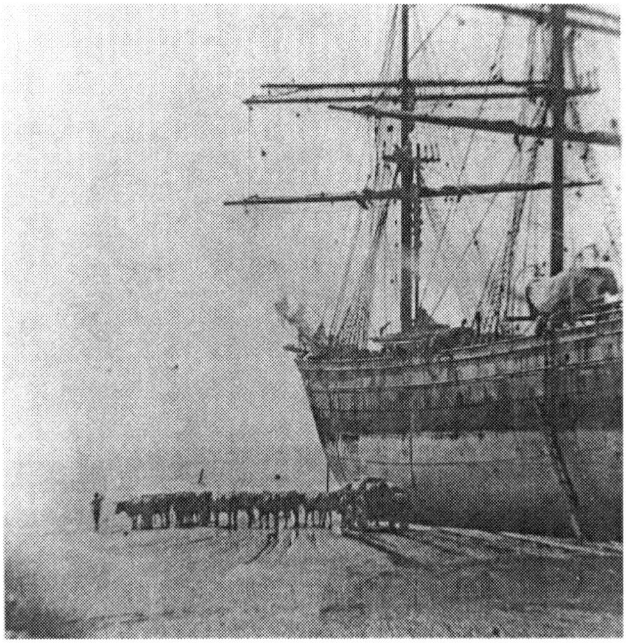

Figure 19. The barque *Arabella* loading at Condon near Port Hedland on the northwest coast (Hick Family).

Eighty three of the intended 100 tons of lead ore were loaded onto *Xantho* from small boats and this cargo was then topped with wool and whale oil from the nearby district and bay whaling establishments (Trenaman, 1934). *Xantho* left Port Gregory for Champion Bay on the night of November 16, heading into a strong southeasterly breeze and a heavy sea. Then it sank beneath Broadhurst and his crew in the manner described in the following chapter.

Chapter 3

Xantho–The transformations

3. A WEALTH OF INFORMATION

Xantho was built in 1848 as a paddle steamer by the Denny Shipbuilding Company, which operated out of Dumbarton in Scotland (Denny and Brothers, 1932). It was the eighth steamer and only the 22nd vessel built by the famous and well-documented Denny Brothers. PS *Xantho* was propelled by a 60 horsepower (hp), steeple engine and it was powered by steam from a tubular boiler, most likely of a rectangular form (Lyon, 1975:118).

3.1 The specifications: Construction as a paddle steamer

A great deal of information about PS *Xantho* exists. The original builder's contract and specifications were found in the Moor Collection of Denny Shipbuilding Company documents and the latter appear in Table 2 below. The contract showed that it had hull plate thickness a maximum of 5/16 of an inch (c. 8 mm) thick, with other scantlings similar to those recommended by Grantham for equivalent river steamers

(1859:186–187). He suggested hull plating at a maximum of 4/16–5/16 of an inch (6–8 mm), for example. When Lloyd's tables for vessel intended for use at sea are compared with the builder's contract for *Xantho* it indicates that it was lightly built in comparison to an equivalent seagoing steamer. Its hull, below the waterline, was over 1/16 inch (1.5 mm) less than that specified if it were to receive even a six-year certificate for use at sea. Its watertight bulkheads, at 1/8 inch (3 mm) thick, were only half the thickness required by Lloyd's for service at sea.

Equally significant, the *Xantho* frame spacings at 21 inches (533 mm), center to center, were greater than what Grantham suggested for river steamers and what Lloyd's required for service at sea. Both were uniformly set at 18 inches (457 mm).

Table 2. A summary of the specifications for the PS *Xantho* (transcribed from originals in the Moor Collection by D. I. Moor)

Length between perpendiculars: 101.3 feet (30.8 m)
Length overall: 121 feet (36.8 m)
Breadth of Beam: 17.6 feet (5.3 m)
Depth of Hold: 8.4 feet (2.5 m)
Keel: Of bar iron, 3 inches x 1 inch [in section] (76.2 mm x 25.4 mm)
Frames: For 40 feet (12m) in midships, 3 x 2 x 5/16 [inch] (c. 8 mm) angle iron. Fore and abaft [aft] that 3 x 2 x 1/4 [inch] (c. 6 mm) to be placed 21 inches (533 mm) from centre to centre.
Floors: Of plates, 9 inches (228 mm) deep by 3/16 inch (c. 5mm) thick with 2 1/2 (c. 64 mm) x 2 1/2 x 3/16 inch angle iron, rivetted on top edge, with extra strength of floors for fastening engine.
Plates: Bottom to 2 feet (610 mm) waterline 5/16 inch from 2 to 5 feet (1524 mm) 1/4 inch to gunwale 3/16 inch to be overlapped longitudinally with flush butts and rivets.
Stringers: Of 2 1/2 x 2 1/2 x 1/4 inch angle iron and plates 12 inches (c. 305 mm) broad x 3/16 inch thick the angle iron to be rivetted to the outside plates and stringers.
Bulkheads: To have three of these full depth of vessel, all of 1/8 inch (3 mm) plates. To be stiffened with angle iron and made perfectly watertight. The plates to be all of one length, and neatly rivetted with snap-headed rivets.
Coal Bunkers: Same thickness, size as required.
Deck Beams: To have one on every alternate frame of angle iron 4 x 2 x 1/4 inches, to be single kneed with triangular plate knees, 12 inches long in the arm by 5/16 inch thick, with 3 rivets in each arm, and rivetted to frame with one 3/4 rivet in each beam end.
Paddle Beams: To be framed of plates 12 inches deep, with 6 x 3 x 3/8 inch angle iron, rivetted back to back.
Decks: Deck plank of Quebec yellow pine, 5 x 2 1/4 inch tapered

Holds: To have hold fore and abaft engine and boiler space size as may be required. Two cargo derricks for use of holds, and a small winch to each derrick capable of lifting 1 1/2 tons.
Sails: Foresail or square sail, mainsail and jib.
Cabins: Main cabin and steerage to be finished complete, in a neat but plain manner. To have haircloth sofas in main cabin and in captain and mates room.
Carving: Figurehead and trail boards, trail boards to be hatched with gold.
Cooking Apparatus: To cook for 8 men.
Paintings: As iron boats usually are done.
Sundries: 2 anchors and 2 chain cables. 2 cork fenders, 1 ensign, 1 Union Jack, 1 burgee and ferry flag, 1 long sweeping broom, one deck scraper, 3 brooms, 1 paint scrub and mop, 2 long brushes for funnel, 2 deck lanterns, 2 holly stones, bell and belfrey, 4 wooden fenders with hooks and chains, passenger gangways with ladders, 2 hand poles and limber chains. To have an iron knee inside on gangway stanchions, hawsers and warps, water cask, 3 pails, axe and saw. Water closet in main cabin, and one in steerage–cabin store, tables and mirrors, 12 camp stools, life buoys according to act, compass to be adjusted,
Engine: To have a 60 horsepower Staple [sic] engine, with a tubular boiler, capable of generating a sufficient quantity of steam for the same. Diameter of cylinder 43 inch. The engine and boiler to be upheld for 6 months by the contractors in the event of materials or workmanship giving way.

Note: In ships' registers and documents of this period, length is expressed in feet and tenths of feet. Where vessels were measured in tons under mid nineteenth-century tonnage rules measurement fractions of 100 can also appear. (See MacGregor, 1993: 283–4 for a succinct précis of tonnage and tonnage rules.)

When a vessel or its machinery is assessed in terms such as horsepower, or built to feet and inches, it is customary to quote the original figures and where necessary to provide a metric equivalent, for often the former shows a pattern that is not otherwise evident in the metric conversion. For example the 21-inch frame spacings on *Xantho* have a metric equivalent of 533.4 mm.

Finally, as one tonne = 1000kg= 0.984 tons, i.e., they are almost identical, tonnage is left as per the original throughout this work and the metric equivalent is not provided.

Xantho's total building costs were £3,270, equally divided between the machinery and hull (Lyon, 1975:118). The name *Xantho* may have been derived from Greek mythology (Xanthus, a river on the plain of Troy) or from the Greek root *Xantho* (yellow). This could refer to the use of yellow pine (*pinus strobus*) on the deck timbers. This North American softwood is recognized for its shipbuilding qualities, especially its stability, ease of working, and limited shrinkage in drying, although it was not resistant to rot (Bramwell and Palmer, 1979). The ship's register (Port of Anstruther, Certificate of Registry, 41 of 1848) shows that the engine room was located amidships, resulting in two cargo holds, one fore and the other aft of the machinery space, each served by separate winches and other attendant machinery. It had one deck and was schooner-rigged with two masts.

Xantho's first owners were the Anstruther and Leith Steamship Company and it was used in this period as a pleasure steamer operating between Leith and Aberdour across the Firth of Forth, near Edinburgh. After 12 years in this service, *Xantho* was sold and transferred to Scarborough. Later, in July 1864, it was sold again, and its register transferred to Wick in northern Scotland, where according to the Mercantile Navy list of that year, it was permitted to take excursions to sea (Henderson and Henderson, 1988). In the following year it is described in the *Glasgow Herald* (June 7, 1865), as a "smart iron, passenger-cargo, paddle steamer." Thus, *Xantho* had an apparently uneventful career as a paddle steamer, operating first in relatively sheltered waters for a period of 16 years and then at sea around Scotland for a further 6 years.

Plans, photographs, or contemporary illustrations of *Xantho* have not been found. Mentioned in the specifications for the building of *Xantho*, however, is Denny's iron paddle steamer *Loch Lomond*. Having been built in 1845 it is expected to bear some resemblance to the *Xantho*, which was comparable in length. In contrast to *Loch Lomond*, the PS *Xantho* was rigged as a topsail schooner and was capable of carrying sails either as an assistant to the steam engine or as a substitute where conditions, or operating parameters, required.

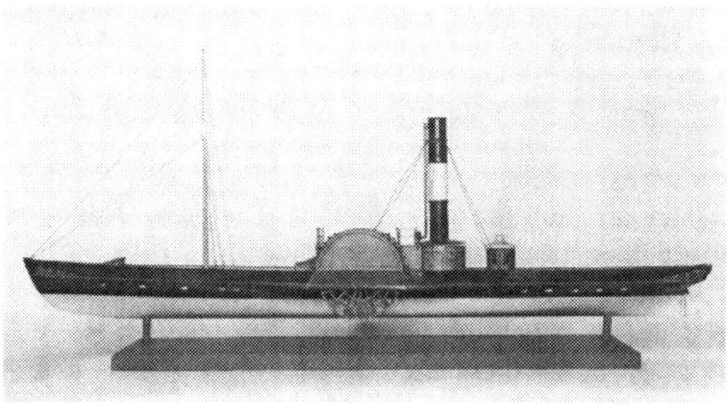

Figure 20. A side view of paddle steamer *Loch Lomond*, showing the hull configuration (Glasgow Museums and Art Galleries).

3.2 The first transformation

In dealing with the transformation of an archaeological site and its assemblages, an important distinction is made between changes that occur as a result of human and natural actions. These Michael Schiffer referred to as "cultural formation processes" and "natural formation processes" (1976:12–19). In applying these concepts to an entire ship like *Xantho*, it can be seen that both cultural and natural transformation processes occur independently or together over the active life of a vessel and then following its abandonment. If a vessel is wrecked these effects can be pre- and postdepositional. In the first instance, these can be corrosion, damage at sea, refit, abandonment behavior, etc., and in the second, examples are salvage and natural processes at the wreck or on the seabed.

The cultural transformations to the *Xantho*'s form wrought by its owners and salvors are reasonably well documented. Some of these were quite dramatic: altering the vessel, not through periodic refit involving repairs or maintenance of the hull but by completely changing its configuration and physical characteristics and then by altering its operating and economic context.

3.2.1 Xantho: A hybrid refugee from the scrap heap

In 1871, *Xantho* was sold to a "metal merchant," Robert Stewart of Glasgow. Seeing an opportunity for profit, he did not scrap it but refitted the ship and altered it from paddle to screw propulsion. *Xantho's* stern and figurehead were also altered and the ship was lengthened to 116.3 feet (35.4m.). The cumbersome paddle engines were replaced with what was initially described in register documents as a 30 hp horizontal engine built in 1861 by John Penn and Son (Port of Glasgow, Certificate of Registry, 61 of 1871). The error was later corrected to 60 hp (See Figure 21 below.)

In making the transition from a large paddle engine housed around midships to a compact horizontal screw engine housed aft, the engine room length was reduced from 32 feet (9.7 m) long to 23.1 feet (7.04 m). This represented a considerable savings in space, with a resultant gain in cargo or coal carrying-capacity. These alterations resulted in the relocation of all the machinery (including the pumps) aft.

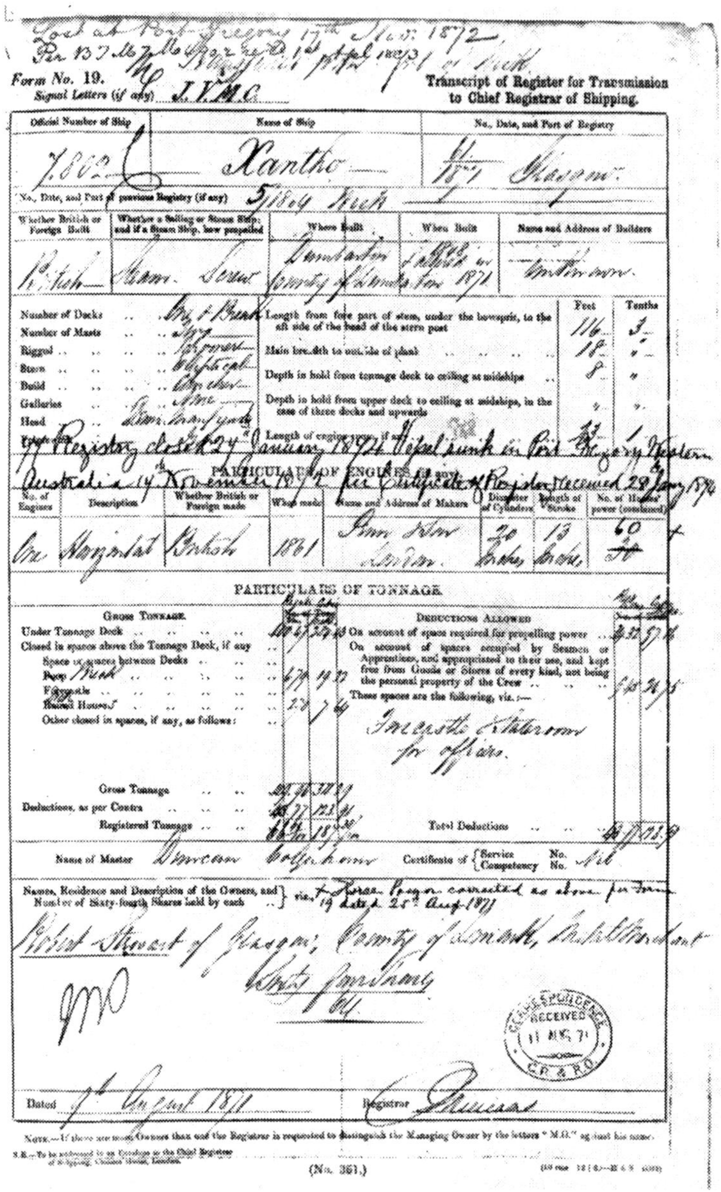
Form No. 19.
Signal Letters (if any) J.V.M.C.
Transcript of Register for Transmission to Chief Registrar of Shipping.

Official Number of Ship	Name of Ship	No., Date, and Port of Registry
7.802	Xantho	61 / 1871 Glasgow

No., Date, and Port of previous Registry (if any) 5/1864 Leith

Whether British or Foreign Built	Whether a Sailing or Steam Ship; and if a Steam Ship, how propelled	Where Built	When Built	Name and Address of Builders
British	Steam. Screw	Dumbarton County of Dumbarton	1848 altered in 1871	Unknown

Number of Decks
Number of Masts
Rigged
Stern
Build
Galleries
Head

	Feet	Tenths
Length from fore part of stem, under the bowsprit, to the aft side of the head of the stern post	116	3
Main breadth to outside of plank	17	
Depth in hold from tonnage deck to ceiling at midships	8	
Depth in hold from upper deck to ceiling at midships, in the case of three decks and upwards		
Length of engine room, if any		

PARTICULARS OF ENGINES

No. of Engines	Description	Whether British or Foreign made	When made	Name and Address of Makers	Diameter of Cylinders	Length of Stroke	No. of Horses power (combined)
One	Horizontal	British	1861	Penn & Son London	20 Inches	13 Inches	60

PARTICULARS OF TONNAGE.

Gross Tonnage.
Under Tonnage Deck
Closed in spaces above the Tonnage Deck, if any
Space or spaces between Decks
Poop
Forecastle
Round Houses
Other closed in spaces, if any, as follows:
Gross Tonnage
Deductions, as per Contra
Registered Tonnage

Deductions Allowed
On account of space required for propelling power
On account of spaces occupied by Seamen or Apprentices, and appropriated to their use, and kept free from Goods or Stores of every kind, not being the personal property of the Crew
These spaces are the following, viz.:—
Forecastle & Stateroom for officers.
Total Deductions

Name of Master Duncan
Certificate of {Service / Competency} No. Nil

Names, Residence and Description of the Owners, and Number of Sixty-fourth Shares held by each
Robert Stewart of Glasgow, County of Lanark
Sixty four Shares

RECEIVED 11 AUG 71

Dated 9th August 1871
Registrar

Note.—If there are more Owners than one the Registrar is requested to distinguish the Managing Owner by the letters "M.O." against his name.

(No. 361.)

Figure 21. The 1871 *Xantho* register (Port of Glasgow, 61/1871, PRO, London)

By relocating the cargo space forward, all the holds could be serviced by one deck winch. The economies of space, time, and labor following from such a rearrangement could have resulted in an otherwise commercially unattractive vessel appearing quite viable. Through this particular transformation process, a redundant artefact (the *Xantho*) was modified and reused, rather than being discarded or being broken up in order to retrieve useful materials, as was often the case. Some of these alterations are manifest in the material record. It is interesting to note under the heading "Particulars of Tonnage" that the figures shown on the register appear in both tons and cubic meters (e.g., a gross tonnage of 109.96 tons or 311.29 cubic meters and a registered tonnage of 66.79 tons and 187.38 cubic meters). This was apparently with an eye to the continent, where the metric system was in place, especially in France. Here is an example of an object of little use to one group being presented in such a way to appear attractive to another and then being offered for trade. In its late 1871 configuration, awaiting a prospective buyer, *Xantho* may indeed have been an attractive proposition for operations on rivers or sheltered European waters. Instead, the ship was taken far out of its original European context for use in the northeast part of the Indian Ocean and on the Western Australian coast.

3.3 The loss of *Xantho*: The second transformation

The next transformation of *Xantho*, i.e., the sinking of the ship appears to have been a product of both old age and incontinence (a natural transformation process) followed by mismanagement and overloading (a cultural transformation process).

The sinking was described in detail at a subsequent court of inquiry and much of it is directly relevant to an understanding of the material remains. The pilot stated that he would not have taken the vessel from port in the trim that it was in, as it had no chance to rise to the sea, for example. As it was originally designed and licensed for use in inland waters, *Xantho* most likely had bows common to river and lake steamers; i.e., without what is termed *flare* or the ability to thrust aside or rise above an oncoming wave or swell. As a result, the bows would have tended to bury into the sea and most of the water would have surged onboard instead of being hurled aside, as was usually the case with a ship built for open water. The deck timbers had also opened up in the semi-tropical heat (see following) and thus the vessel's design may have been unsuitable for the persistent, short seas of the midwest coast.

The first mate then provided further details on the course of events and in doing so provided us with an even greater amount of archival material about the ship.

Table 3. Excerpts from the evidence given at the court of inquiry by the first mate. (Reproduced from: Correspondence relating to the loss of the steamship *Xantho*, and evidence at the Court of Inquiry, 1872. CSO, 727/233-275, State Archive of Western Australia (SAWA)

. . . We shipped . . . 83 tons of lead ore. We left . . . for Champion Bay [Geraldton] on the night of the 16th at 9.00 pm. The wind was SE. It was a strong breeze and a heavy sea. . . . The cargo was stowed under the captain's direction. We expected to take 100 tons on board. . . . Had the 100 tons been taken on board a considerable proportion of the deadweight would have been in the after part of the ship. The Captain said he would only take 1 boat load more. Part of the cargo of ore was removed to the after part of the vessel. When we had finished taking cargo on board the *Xantho* she was about 5 or 6 inches by the head, her usual loading trim is about 2 inches by the stern. She was then 7 or 8 inches out of her proper trim. . . . It was my watch until 12 that night. During my watch the vessel was not taking in water more than I had before seen her do in a head sea. I was down in the fore compartment about half past eleven I did not notice any water in the fore compartment. The *Xantho*'s decks leaked a good deal. I went to see if any of the crew were asleep in the forepart of the ship as [indecipherable]. . . . I was relieved at twelve by the Captain. I didn't make any report to the Captain as to the water the vessel was taking in. I did not consider it excessive. I had before seen her taking as much water. . . . The fore hatch was battened down. The swinging doors of the forecastle were closed and the side also. I went to my berth on being relieved. At five minutes past I was called by the Captain. He told me he wanted me on deck to look after the hands as the fore part of the ship was full of water. I went on deck and passed the Captain at the wheel. He told me to go forward as the ship was in a sinking position. The *Xantho* had two watertight compartments in her. On going forward I found the whole of the forepart of the ship under water it being level with the combings of the fore hatch, there was as near as possible a difference of elevation between the stem and the stern of 7 feet. Part of the lead was thrown overboard to lighten the vessel. . . . I went aft and recommended to the Captain to return to port. He put the stern around. After the vessel was put around I observed he was going out to sea and I requested the pilot to go to the helm and steer for the port for the purpose of saving lives, which he did. Until then the Captain had the helm in his own hands. The engine pumps were going but were of no use, the water being all forward. There were no pumps in the forepart of the ship. We reached Port Gregory at a quarter to four on the 17th. We went in the Hero Passage. She took the ground about 10 minutes after entering the passage when abreast the Gold Digger Passage. The water then began to go aft. It was not more than 15 or 20 minutes after the water began to run aft that the fires in the engine room were put out by it. The vessel then settled down. The pumps were then useless. . . . Attempts were made to free her by bailing but it had no effect on her. . . . At any rate the bailing had not the slightest effect in resolving the quantity. I cannot account how the water got from the fore compartment. . . . the Captain...gave no orders. He did not appear to be competent to give orders. He had lost the presence of mind altogether. . . . *Xantho*'s port of registry was Glasgow, she is an iron vessel rigged as a topsail schooner. . . . By the vessel being down by the head the water that came in overall could not get out of the scupper holes in the same way it would had she been in proper trim. The *Xantho* was ashore in the Flying Foam Passage on the voyage from Point Walcott, I did not think she received injury there to account for leakage on the night of the 16th. The fore compartment was sound. The bulkhead of the fore compartment was about 15 ft from the stem.

Captain Denicke commented that when he found that the crew were desperately heaving sacks of lead ore overboard on the race back to shore, Broadhurst called a halt, stating that he preferred to "save the cargo rather than the ship" (Minutes of the Inquiry,1872). Under examination on his role in the vessel's demise, Broadhurst revealed that he had forgotten to renew the ship's insurance. Realizing this about a fortnight before the ship sank, he had sent an urgent letter home to rectify the situation, but it was too late, the mails were so slow and the telegraph to Europe was not connected. They were all exonerated from any blame in the loss of *Xantho* by the inquiry. Its findings were echoed by the local press who claimed that:

> Her hull is weatherbeaten and worn out. . . . The vessel was simply swamped through her unfitness from age, service and other causes to carry the freight with which she was laden (*Herald*, January 25, 1873).

3.3.1 Abandonment behavior and its effect on the material record

Of importance to the archaeological record is the speed and progress of abandonment. The loss of *Xantho* eventually proved not to be a life-threatening event, occurring in shallow water within the approaches to an existing port. As a result, abandonment was not hurried. The crew also had ample opportunity to return to the ship after it had settled on the seabed. All accessible valuable or reusable loose material, and most likely all available personal effects, would have been retrieved.

Abandonment processes range from relatively gentle circumstances, e.g., running onto a submerged sandbank at the entrance to a sheltered port like that described above, to people being unexpectedly and abruptly cast ashore on a hostile coast, or into raging seas. Clearly the archaeological record is markedly affected by these variables and they must be taken into account before conclusions are made about the significance of the remaining assemblage (Gibbs, in prep.).

Broadhurst was keen to salvage ore from the wreck and, shortly after it sank, a close examination of the wreck was made from the surface. The foredeck was three to four meters underwater and the afterdeck about one meter below the surface. The engine room, cabin skylights, and cabin companion had all been washed overboard, as had part of the main deck and the bulwarks. The reports concluded that the vessel was filling with sand and that it was a total wreck and should be sold.

3.4 Breakup and salvage: The third transformation

Once a shipwreck has occurred, the vessel is subject to cultural change through the effect of immediate post wrecking salvage. It is then subjected to natural transformations as it disintegrates on the seabed, firstly by the action of the seas and swell and then by accelerated corrosion.

By December diving apparatus had been obtained and salvage of the gear and equipment within the *Xantho* proved quite successful. As a result, the amount of material landed on the beach and later sold at auction was substantial. This included a complete set of sails with running gear, anchors, 81 fathoms of chain, boat davits, lifebuoys, a barometer, thermometers, salinometers, navigation lights, fenders, a large ship's bell, a portable forge, three compasses, a "patent" log, engine room tools, two clocks, lamps, a telescope, and a 13-foot (4 m) dinghy (*Inquirer*, February 5, 1873). The list indicates that the vessel was stripped of everything that could be obtained, including material from the engine room. This apparently left the ship virtually an empty shell, bar the machinery that was fixed to the hull, items buried in the bilge, and much of the lead ore.

Where undertaken by the owners, insurers, or their agents, as in the instance above, the recovery process is what could be termed *primary salvage*. This allows a distinction to be made between the processes of salvage that occur before and after the wreck is totally and finally abandoned by its original owners, the insurers, or their salvage agents. Salvage undertaken by individuals or groups other than those directly involved at the time one could term *secondary salvage* and it generally occurs after an abandoned site is relocated and then salvaged by people without a "primary" interest in the site.

In furthering the process of primary salvage at *Xantho*, the captain eventually proceeded with the sale of the wreck in order to obtain funds with which to pay the crew. The auction was poorly attended and it was subsequently described as a "complete sacrifice." Items such as the dinghy on the beach at Port Gregory fetched only £1. As a result, the total sum raised was only £180. Of significance was the fact that the hull and engines were sold as one lot, fetching £110. Interestingly, in the face of a complaint from Eliza Broadhurst (while Charles was on his way back to the wreck), the purchaser refused to pay for the engines and hull until he was guaranteed that the sale was authorized by Broadhurst himself. There was considerable confusion on this issue, for Broadhurst had apparently not yet relinquished ownership. In March and April 1873, for example, he was still calling for tenders with the aim of raising the steamer (*Inquirer*, March 19, 1873; *Herald*, April 5, 1873). That the wreck was still salvable and accessible is of importance in understanding the material record. Equally of significance is the fact that the purchaser did not complete the deal or exercise his options as expected. As a result, the engine and hull remained on the site.

Strangely, one of the reasons for this fortuitous outcome was Broadhurst's poor labor relations. When the ship sank, Broadhurst did not pay his men off in the belief that, as the vessel was salvable, they were still in his employ. With this understanding he went off south to recover what he could of his faltering business empire.

On the other hand, the captain and the crew, though left without instructions or money, were of the belief that as the vessel had sunk they were entitled to be discharged and to be paid their dues. They pursued Broadhurst down and then back up the coast, seeking their arrears in wages. In the meantime four Malay divers were left destitute, wandering the streets of Geraldton. The matter became a great public scandal in which Broadhurst was roundly criticized. In taking up the case of the destitute sailors, the local press urged the sale of the wreck and its machinery to cover the arrears.

The salvage, the lifting of the decking, the loss of the skylights, and the inevitable removal of the valuable masts, spars, and rigging, indicate to the author that the hypothetical *Xantho* shown in Figure 16 could have been transformed within a few months of its loss as shown below.

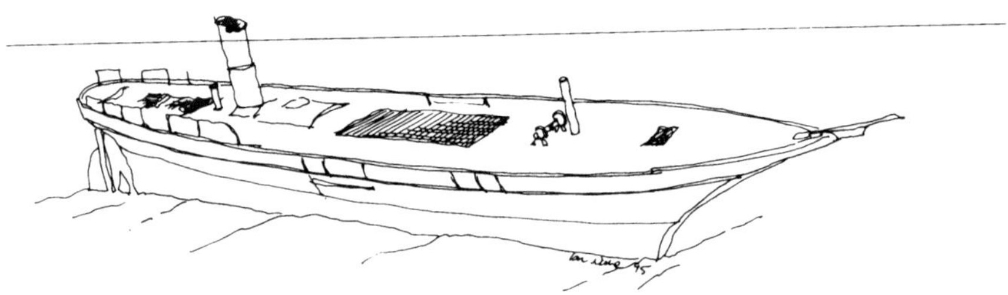

Figure 22. A sketch representing the possible appearance of the ship a few months after its loss (Ian Warne).

3.5 Another contemporary record of *Xantho*?

Because it amounted to little in their eyes, the Broadhurst family appear not to have produced or kept any illustrations of one of their greatest failures, the ill-fated *Xantho.* Nor did other Europeans, presumably for the same reasons. In contrast, there are those illustrations attributed to indigenous people (e.g., Figure 18) and it also appears that another may have been produced by one of the Malay divers Broadhurst brought to Australia on *Xantho*. Some of these men he was known to have abandoned at Geraldton and later at Shark Bay. (See the section *Broadhurst Revisited* in Chapter 6).

Around 1917, some 45 years after the loss of *Xantho*, a former Shark Bay pearler named Sammy Malay is reputed to have joined an Aboriginal group at a soak near Walga Rock inland from the mouth of the Murchison River (Figure 13). Soon after he arrived there a ship painting appeared in a nearby cave (S. Gratte to P. Playford, January 9, 1997).

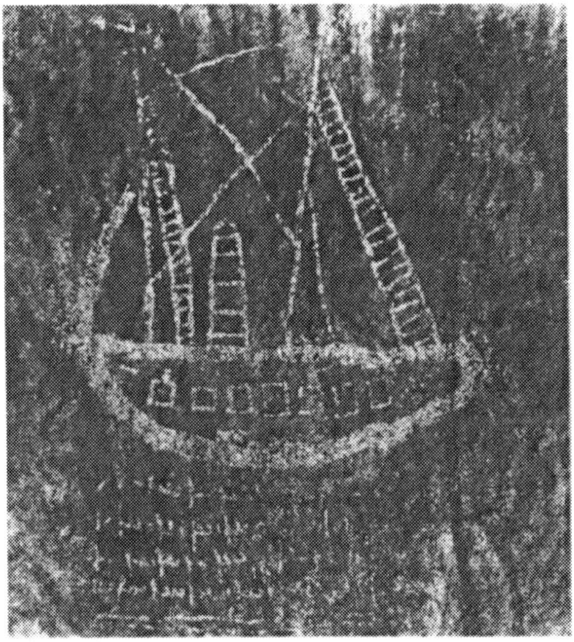

Figure 23. The Walga Rock painting and a steamer of *Xantho*'s vintage and description compared (photograph Phillip Playford; illustration M. McCarthy and Ian Warne).[3]

[3] The Walga Rock painting is reproduced by permission of the traditional owners of the area through Dr. Playford and Dr. Ian Crawford. The false gun ports in the lower illustration are conjectural, but were common on iron ships of the period, especially those operating in waters where pirates or other threats were expected. There are numerous contemporary examples.

Until recently, most believed the painting to be a record of a heavily armed Dutch East Indiaman, the *Zuytdorp* (1702–1711) that was lost on the coast (e.g., Playford, 1996). Many, including one prehistorian now think it was the *Xantho* (I. Crawford, personal communication, December, 1996).

3.6 Xantho: A navigational hazard

It also appears that the submerged hulk at the entrance to Port Gregory was quickly forgotten by all except those navigating the narrow channel in which it lay. In 1875, for example, the area was surveyed and the site was marked with the notation, "submerged wreck." Commander W. E Archdeacon RN, does not name the vessel, but he describes the location exactly in his log:

> One third of a cable [c. 70 m] off the point a small coasting steam vessel foundered and her remains not having been removed it is probable the point will eventually work out to it (Archdeacon, 1879:16.)

Soon it became lost to living memory. Of equal importance was Archdeacon's belief that the point (or sandbar) would "work out" to the wreck and that it would, by inference, become engulfed in it. From Archdeacon's time on, we know little of what occurred at the site, but on days of complete calm and clear water sailors could not have missed seeing the wreck as they tacked into Port Gregory. Lying so close to the channel, it was also a distinct and quite noticeable hazard to navigation, with the boiler and upper-works barely three meters below the surface.

Many nineteenth-century pearling craft en-route to Fremantle or the north would have utilized Port Gregory, the only shelter on the long stretch of coast from Shark Bay to Geraldton. With divers onboard, they may have investigated the site, removing anything that still appeared useful to them.

Indonesian trochus shell divers operating on the Australian coast and off-shore islands, for example, are known to have regularly visited wrecks, reducing them to sterile sites (McCarthy, 1991). See the divers alongside the 1880s iron barque *Ann Millicent* in Figure 39, for example. Given that it was such an obvious obstruction at the entrance to the harbor, pearl divers traveling through Port Gregory can be expected to have done the same.

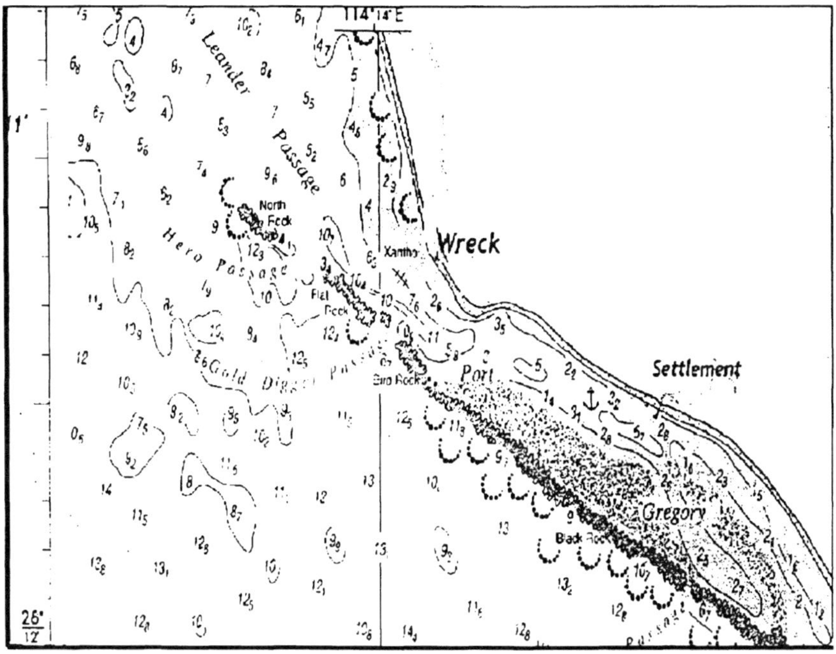

Figure 24. Port Gregory, in modern times, showing the location of an obvious navigation hazard (the *Xantho*) at the entrance to the port (Department of Land Administration, Western Australia).

In the early twentieth-century, salt was harvested from inland lakes and shipped from the port and it appears that we are fortunate in having anything at all to examine at the site. When the SS *Kurnalpi* called into Port Gregory in February 1918 to take on a cargo, the master requested that the wreck be removed as it constituted a hazard to his navigation. Fortunately, it was decided to mark it with a permanent beacon instead of destroying it with explosives, as was usually the case. (Suckling n.d.). Though the nature and form of the beacon on the wreck is not known, it is expected that the marker would have been held in place by chains fastened to a portion of the wreck itself, further hastening its destruction in times of bad weather.

In the interim, disintegration would have continued as a result of natural processes. Experience shows that, if subjected to wave and swell action from beneath, e.g., through open hatches of breaches in the hull, wooden decking quickly disintegrates. Then the underlying iron deck-beams flexing with each heavy swell gradually weaken to the point where they fail. This leaves the sides of the vessel without support and totally vulnerable to the seas (e.g., the SS *Macedon* shown in Figure 35). As they corrode further, or are exposed to heavy swells across the hull, the sides collapse, especially around the cargo holds. The direction in which they fall is dependent on the direction of the seas at the time of near-collapse, the current, and the angle of heel of the hull itself.

In contrast, the stronger and better-supported areas around the boiler, coal bunkers, stokehold, and engine room often form a unit and remain intact for a considerably longer period (e.g., the SS *Colac*, Figure 3). The bow and stern, which are immensely strong structures, still remain upright and they often remain so for some time.

These phenomena have been observed in contemporary photographs and illustrations of ships in similar situations and they are now believed to be a common process, notwithstanding salvage or other cultural transformations, such as the effect of war. This, somewhat surprisingly, is an issue which had to be considered at the *Xantho* site, for the area was randomly shelled by a Japanese submarine in 1943 (Hashimoto, n.d.). Investigations subsequently showed that the site was not damaged (MacDonald, 1994).

The author's conceptualization of the appearance of *Xantho* as it began to collapse is represented by the team's project artist in the illustration below. This hypothetical impression (and those at Figures 16 and 22) are not scientifically based and there are numerous assumptions inherent in the production of any illustration purporting to show how *Xantho* would have appeared in the years between its abandonment and its relocation and inspection in modern times. These revolve around the notion that there are recognizable commonalities in iron ship disintegration—an issue that will be discussed at length in Chapter 6.

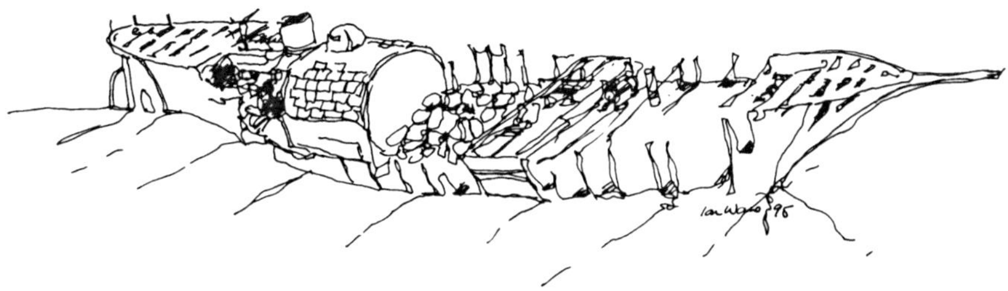

Figure 25. An impression of the appearance of the *Xantho* some years after its loss (Ian Warne).

Chapter 4

The wreck examined

Conservators and archaeologists on the seabed

4. PROCESSES IN WRECK SITE ANALYSIS

Nearly 100 years after *Xantho* was lost, a member of the Underwater Explorer's Club (UEC) of Fremantle found the site while perusing early charts and modern aerial photographs (*Underwater Explorers' Club Newsletter*, October 16, 1966). Little was done about the report and this was probably fortuitous, for when the wreck of the SS *Georgette* was found in the same period, everything possible was removed and the engine was dismantled for souvenirs, leaving it broken on the seabed (Figure 10). In an attempt to remove the propeller with explosives, it was blown to pieces and though the spare was dragged ashore, it was lost in the sand. In mitigation, it must be noted that such actions were the norm at the time.

4.1 The discovery and inspection of the site

Research conducted for the Western Australian Museum's "colonial wreck program" (Henderson, 1977) indicated that *Xantho* lay close to Port Gregory. Having been requested to locate it as part of that program, avocational shipwreck enthusiasts from the Maritime Archaeological Association of Western Australia were shown the site by local fishermen. The Association's report, of an iron wreck at the entrance to Port Gregory, with a boiler, engine, and cargo of lead ore, left no doubt that it was the SS *Xantho* (Hall, Hill, and Warne, 1979). It was subsequently inspected by the museum, some materials were raised, and a report was filed (Sledge, 1979).

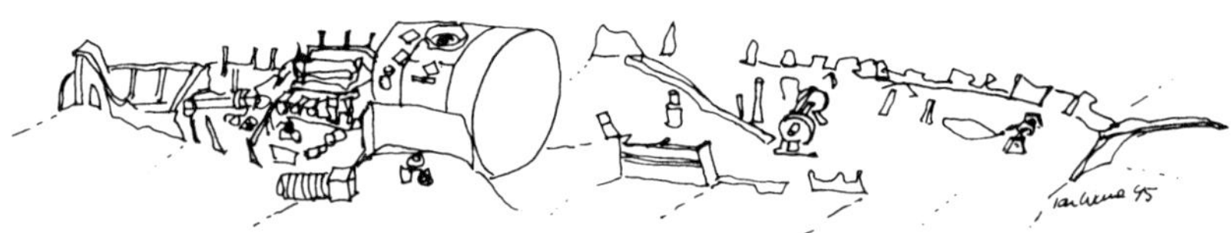

Figure 26. A sketch of the SS *Xantho* as found in 1979 (Ian Warne).

In January 1980 the association noted an intact boiler gauge near the aft starboard corner of the boiler. In October they returned to find it gone. Concern was expressed and they recommended to the museum that all loose materials be removed as soon as possible (Totty, 1982). The author was then requested to conduct a detailed survey and surface recovery in order to remove the temptation and to ascertain the extent of the deposit through test excavation (Henderson, 1988). *Xantho* was to be the first iron-hulled vessel excavated under the museum's colonial wreck program and before embarking on the project a literature search on iron and steamship excavations was conducted.

4.2 Early iron and steamship archaeology examined

In the 1970s, four years before *Xantho* was found, the editors of *The International Journal of Nautical Archaeology* had warmly welcomed Gordon Watts's report on the location of the *USS Monitor* (1862–1862). It was described as an "important paper", heralding the acceptance of iron and steamship studies in maritime archaeology generally (Watts, 1975:171). This work and those that had preceded it (e.g., Baker, *et al.*, 1969; Edington, 1978; Edington *et al.*, 1978; Friend, 1978; Watts, 1972; Watts and Bright, 1973) were contemporaries of a number of important American wooden steamship studies. Examples are the ill-fated USS *Cairo* (1861–1862) recovery program (Bearrs, 1963) and the removal of the machinery from the "propeller", or screw steamer, *Indiana* (1848–1858) in 1979 (Jacobs, 1979).

There were also numerous iron ship restoration projects underway when the *Xantho* project began (e.g., Brouwer, 1985; Corlett, 1970; Envig, 1984; Wall, 1979; Reynard, 1979) and a detailed study on the sinking of USS *Maine* (1895–1898) was relevant (Rickover, 1976). The 1980s revitalization of the *Cairo* program (McGrath, 1981) and the part-excavation of the sidewheel cotton packet *Black Cloud* (1864–1873) by students of George Bass (Adams, 1980) were two other important contemporary studies.

Much of the American steamboat research (e.g., Simmons, 1988), and other studies contemporary to the *Xantho* project, such as Canadian research into the SS *John Fraser* (1883–1893) and the survey of the wooden-hulled propeller *Conestoga* (1919–1922) were published after the *Xantho* program commenced, however (e.g., VandenHazel, 1987; Gregory, 1984). A number of contemporary thematic studies, e.g., a reassessment of the California Gold Rush through maritime archaeology and the subject of "blockade running" (e.g., Bright 1985; Delgado, 1983; Wise, 1985) were also produced or received after the *Xantho* program began.

The *USS Monitor* and the steel-hulled USS *Arizona* (1915–1941) in America and the TSS *John Penn* (1867–1869) study in New South Wales (Australia) were also contemporary projects, directly relevant to the *Xantho* program. Again details were not available in 1983 (John Penn Team, 1984; Murphy, 1987; Lenihan, 1989; Lorimer, 1988; Watts, 1987). As a result there were few "ready theoretical formulations" (cf. Renfrew 1982:3) to prescribe an obvious way to proceed in iron/steel and steamship archaeology when the *Xantho* program commenced that year. The isolation was keenly felt, especially given Muckelroy's earlier disclaimers about the value of post-1800 research.

4.3 A predisturbance study: New direction in 1983

Expedition aims and a request for funding the *Xantho* program to a budget of $3,700 AUS were subsequently presented. In the submission, the problem of effectively recording an iron steamship site with substantial relief, such as boilers and engines, was identified as an important consideration. As a result, wreck inspection technique, recording method, and excavation philosophies were re-examined. An objective was also to comment on the research potential of the hull, the propulsion system, and cargo remains, and to assess and report on physical conditions affecting the wreck and its future stability. In doing so, the intention was to make no *a priori* assumptions about the *Xantho* site, how it should be treated, or how it behaved in an underwater environment in comparison to the wooden-hulled wreck. As a result it was decided that a full predisturbance survey had to be carried out prior to the main recording/excavation program.

As indicated by the literature search, there was unfortunately little precedent on which to base the predisturbance work. All previous shipwreck corrosion studies were undertaken post- or mid-excavation, e.g., the in-laboratory examination of materials recovered from the ironclad USS *Tecumseh* (1863–1864) by Baker *et al.* (1969) and the postdisturbance examination of materials at a number of wooden-hulled wrecks off the coast of Western Australia (MacLeod, 1981; North, 1982).

Just prior to the study commencing, guidelines for conducting a co-ordinated wreck inspection program were published in *The International Journal of Nautical Archaeology*, however (McCarthy, 1982). This made provision for the examination of both wooden-hulled and iron-hulled ships. Included at the suggestion of the Western Australian Museum's conservators (Neil North, James Pang, and Ian MacLeod) were a number of artefact management and site-related parameters. These and their relevance to the study of shipwrecks are described in the following table. Added to the original, for the purposes of the 1983 *Xantho* study was provision for the measurement of the electrochemical state of its hull, machinery, and other ironwork.

Table 4. A list of biological, physical, and chemical parameters of relevance to the conserving and managing of shipwrecks

a) Temperature: Water temperature is an important variable in determining the rate of marine growth, corrosion of metal objects and the biodeterioration of organic materials.
b) Salinity: Water salinity has a pronounced effect on the stability of metal objects, ceramics, and biological growth, e.g., bacteria and fungi that have a marked effect on the deterioration of wood. Salinity, therefore, is an important factor to be considered and measured.
c) pH and dissolved oxygen content: Both should be measured on site where possible as they are major factors in determining the stability of both inorganic and organic materials.
d) Water movement and purity: Wave action, water movement, and non marine water sources nearby, e.g., rivers, sewers, etc., should be recorded in view of their effect on a–c above.
e) Bottom-type analysis: This variable also needs to be recorded since it has chemical as well as immediate physical effects on the wreck material and on the site itself. Samples of bottom sediment should be taken with a view to analyzing microorganisms present and mainly the sulphate-reducing bacteria content with a view to their effect on organic material and artefacts.
f) Corrosion products and marine concretions: These are best initially retained, where practical, due to their protective coatings, which help prevent damage of artefacts in transit and also for the information conservators can gain from an analysis of the corrosion products and marine concretions. Some marine concretions, however, harden considerably on artefacts such as ceramics, and experience has shown that these are best removed soon after being taken from the site.

As the wreck also lay in very shallow water, precluding the need to have highly experienced divers, it became evident to the author that the most appropriate people to examine these parameters were specialists with experience in the handling of scientific equipment and in interpreting the results. With this self-evident fact and the predicted site conditions in mind, the author (then a diving instructor) trained those who were unable how to dive and to apply their skills to the *Xantho* site. As a result, the *Xantho* study presented conservators with an opportunity to conduct an underwater (as opposed to an in-laboratory study) of long-term site degradation. There was also an opportunity to study biological growth on the site. Clearly, for the results to be most meaningful, the study had to be conducted by the appropriate specialists before the site was transformed in any way; i.e., the work had to be perfomed underwater.

After a short familiarization dive, an assessment of the physical and chemical status and the biological growth on the wreck was conducted by Neil North, assisted by Ian MacLeod and visiting American biologist, Ms. C. J. Beegle. From their perspective, the rationale for the study was that

> very little is known about the rate and manner in which the ship's material decays and corrodes, what type of problems are likely to be encountered when attempting to excavate or raise artefacts, or how to protect any significant sections which have to remain *in situ* after the excavation is completed. Our aim in carrying out this preliminary survey was to collect enough data to answer some of these questions, or at least pinpoint where further work is needed. The type of information which can be obtained from the *Xantho* is applicable, in part, to other marine problems such as the formation of artificial reefs and protection of long-term off-shore facilities (Beegle, MacLeod, and North, 1983:3).

4.3.1 Natural transformation forces analyzed

An examination of the area showed that the prevailing southwest winds and swell had two main effects on the *Xantho* site. One generated a current traveling up the reef system and out over the *Xantho* site from the port to the starboard quarters, the other refracted around reefs into Gold Digger Passage, which lies opposite the site. These seas and swell impinged at right angles to the starboard side of the wreck. There are also sufficient gaps in the reef, opposite and to the north of the wreck, to allow the seas to impact on the *Xantho* site in all conditions except an offshore breeze. The tides are diurnal, in the order of up to 1 meter maximum, and have a direct influence on current, which was strongest just after high tide, and then for an hour or more after (J. Worsley, personal communication, January 1983). The site is also exposed to the full force of the northwest seas and swell that are the destructive precursors to cold fronts which regularly impinge on the coast in winter. An early account of the magnitude of the natural forces in the locality comes from the experiences of the American whale ship *Iris,* which was nearly wrecked at Port Gregory in the winter of 1855. These are described in a contemporary newspaper report, which reads as follows:

> A gale started to blow hard from the north. During it, a strong current swept the *Iris* out towards the open sea stern first, against the wind. Captain Davok had three anchors out, one of which became fouled up with a government mooring buoy. They were all swept away together. . . . the anchors seemed to drift faster than as the vessel as if the whole bottom of the anchorage lifted bodily four ways (*Inquirer*, July 11, 1855).

The conservator's report showed that the wreck lay in a maximum of 4.9 meters of water and that the seawater temperature was 23°C with no thermal gradient. The current, which ran from the port bow across the site to the starboard quarter astern, was approximately 3 knots at its strongest. Visibility was normally good, though the storms that preceded the inspection produced a mass of weed and seaborne grit, which reduced visibility to around 3 meters at best, often falling to less than one meter. The salinity of the water was 37.53 parts per thousand, with dissolved oxygen estimated at 100% saturation. The pH of the water at 23°C was 8.1 and it was found to contain concentrations of lead sulphide downcurrent from the lead ore cargo that were 11,500 times higher than normal seawater (MacLeod, 1992a).

Compared to the nearby barrier reef and benthic communities, the ecosystem of the *Xantho* appeared to be an anomaly. The surrounding areas comprised eelgrass communities with a large herbivore population which feed on the organisms seeking shelter and protection among the eelgrass fronds. Life on the *Xantho,* however, was a tunicate-dominated community, primarily of sedentary filter feeders. These served to camouflage much of the wreck aft of the boiler and possibly to protect the remains from wave action. Also present were tubeworms and a single crinoid.

On the top of the boiler, which was 2.9 meters from the sea surface, encrusting sponges and green algae occurred. There the wave surge was found to be at right angles to the current and approximately 0.5 meters in amplitude on the day of the study. Within the shelter of the remains of the steam dome tunicates were found. These were generally smaller forms of 7–10 cm overall length. A distinct demarcation of growth along the boiler surface was noted at a depth of 3.7 down to 4.2 meters. There, a band of large brown algae occurred. Below this line, large tunicates (10–15 cm overall length), large upright sponges (5–10 cm high), and a few scattered red algae appeared. Above the line, the sea life was associated with that found on the upper surface of the boiler (Beegle, MacLeod and North, 1983).

A thick layer of a rock-hard substance aptly termed *concretion* (cf. North, 1976, 1982) had formed over the boiler, the engine, and all other iron surfaces. On the engine it appeared to be up to 50 mm thick; on the boiler it was less. In some areas, animal matter grew on top of this layer. In comparison to the thick layers found on the iron, the concretion layer on brass and copper fittings was only a few millimeters thick and consisted of a dense white calcareous deposit.

The mechanism for the development of concretion is of significance and is briefly discussed in an ensuing chapter. To deliberately remove more than is absolutely necessary of this layer on a preliminary nondisturbance survey such as that being conducted at *Xantho* is frowned upon today, especially as it exposes the surfaces below to the elements. As a result, the corrosion specialists minimized the impact of their study by clearing the loose plant and animal growth from the area to be examined by hand and then by drilling 6 mm diameter holes through the concretion down to the metal itself. This procedure was performed using a masonry bit and a hand drill, allowing the depth of the concretion layer on each feature to be measured. A platinum electrode connected to a high impedance digital multimeter, housed in an epoxy body, was then inserted into the hole. A reference electrode was placed adjacent to it and the voltage was measured. The procedure allowed the corrosion specialists to ascertain the electrochemical environment and the physical state of the metal beneath. From there they were able to make predictions on its stability. In this manner the corrosion potential of the ironwork and other metallic surfaces were measured across the site.

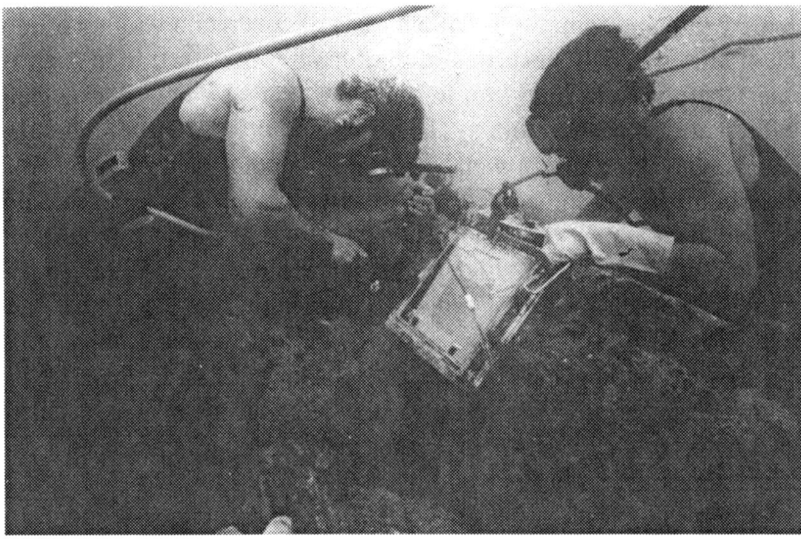

Figure 27. Neil North and Ian MacLeod at work on the engine during the predisturbance survey. MacLeod is using a handdrill; North is holding the multimeter. The electrode is visible in the foreground (M. McCarthy).

4.3.2 Results of the predisturbance survey

Earlier studies had indicated that the rate for underwater corrosion of discrete steel and some iron objects in an anaerobic environment averaged 0.1 mm per year (ranging from 0.02–0.195 mm/year) (La Que, 1975:383–9). These results were supported by a subsequent study, where a mean rate of 0.08 mm per year was recorded after 16 years of measurement (Southwell, Bultman, and Alexander, 1976). In noting these and in having conducted his own studies on corrosion and galvanic coupling, North expected a general corrosion rate of 0.08–0.10 mm/year on *Xantho,* while predicting that it could be expected to vary greatly across the site (N. North, personal communication, January 23, 1984).

The corrosion rates were found to vary considerably across the entire wreck, across the boiler, and even across the engine. During the predisturbance study, for example, the hull plates were found to be extremely thin and some were hollow casts of concretion with no original metal at all, especially upcurrent in the forward, port section of the ship. Further, there was no solid iron left in the winch or windlass, each being in effect hollow concretions. The ironwork on the boiler varied in thickness with some robust metal remaining in parts, especially on its aft face. In comparison, the coppers and brasses appeared to be in excellent condition beneath a very thin layer. Only the engine, drive shaft, propeller, and part of the starboard quarter at the stern appeared to have some solid metal. North then came to the opinion that the engine had a life span of 60 to 100 years at most if left totally undisturbed on the seabed (N. North personal communication April, 1983). He also advised that after this period the engine would be reduced to mere shells of concretion or would collapse under the force of the heavy seas and swell that sometimes affect the site.

In summary, North's specialist team concluded that a recognizable trend to more negative corrosion potentials of cast and wrought iron moving from bow to stern indicated a slower corrosion rate aft. For example, a 168-millivolt difference between the windlass and the engine was the equivalent to the windlass corroding three times faster than that part of the engine measured. The 27-millivolt difference between a hull plate at the bow and the counter astern indicated that the former was corroding 20% faster than the latter. The winch and windlass were in the most exposed environment and their condition appeared to reflect that fact. It was concluded that the wreck and its features, including the hull and engine, were actively degrading and were not expected to last intact much more than another half a century (Beegle, MacLeod, and North, 1983). This was a surprise, for at the time most practitioners assumed that iron wrecks, especially those with considerable relief and apparent structural integrity, would last for a considerable time, both above and below the seabed. Reports on this phase and these understandings were subsequently published (MacLeod, North, and Beagle, 1986; McCarthy, 1986a-d, 1987, 1988a, 1989; MacLeod, 1986, 1989a-b, 1992a).

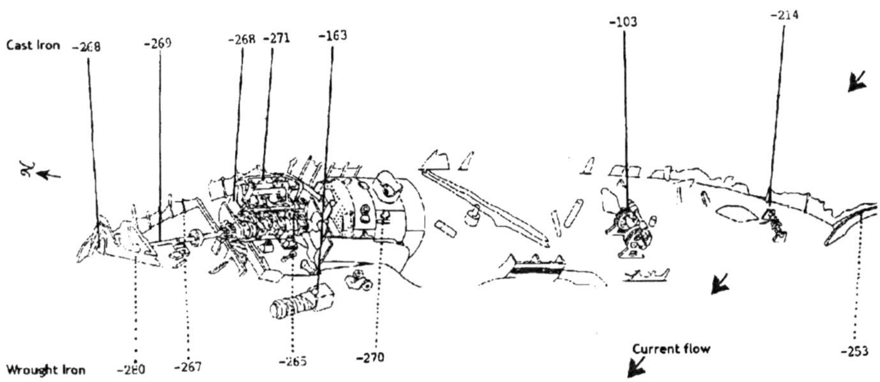

Figure 28. North and MacLeod's corrosion potential measurements at the *Xantho* site. The direction of the current is also shown. Derived from John Riley's isometric projection of the site.

4.4 Anomalous features identified

While escorting the conservators throughout the predisturbance study, it became apparent to the author that the engine was not, as originally reported, lying on its side. This suspicion was confirmed when brass oil cups with lids opening upward toward the surface of the sea were identified. Clearly they would not function on a vertical engine. Further, the area beneath was clear of sediment, allowing its lower parts to be examined more closely and it became evident that the engine was raised off the engine room floor on lateral bearers—it was a horizontal engine. More importantly, two hollow cylinders projected out of the engine block, indicating that the unit was of the trunk engine variety. As indicated earlier, it was a favorite with the Royal Navy from the mid-1840s to the mid-1870s and was a type unexpected outside the naval context (Banbury, 1971).

The identification of the *Xantho* engine as one of the rare trunk engine variety was a complete surprise and an example of one of the problems faced in relying purely on the historical record. The 1871 register (Figure

21) refers to the *Xantho* machinery only as a horizontal engine, for example, for often the engine type is not specified in the registers and in most cases only the barest details are given.

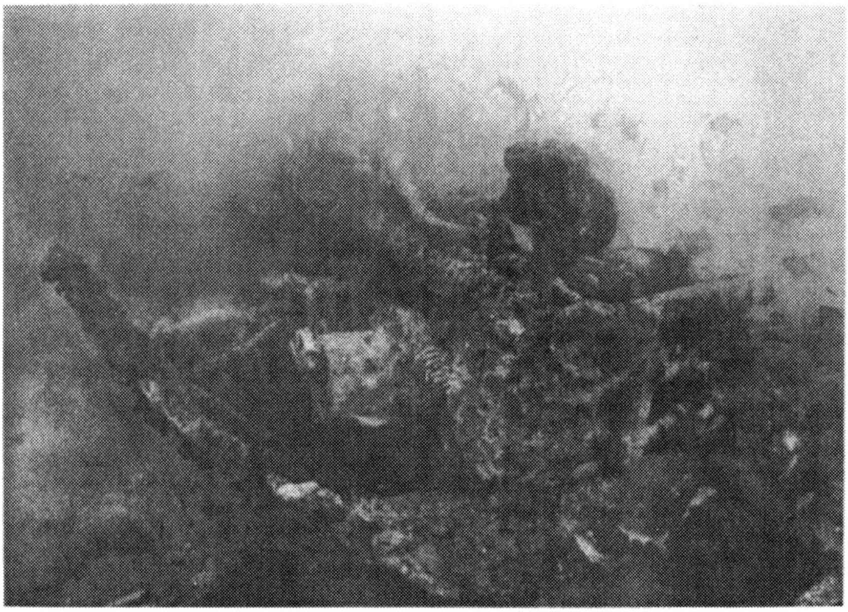

Figure 29. The SS *Xantho* engine, showing the oil cups and the trunks. Note the concreted boiler tube brushes on the right-hand (aft) trunk (M. McCarthy).

As the predisturbance survey continued it became apparent that the engine was even more significant as an engineering artefact in its own right, for *all* of its fittings, copper piping, brass taps, cocks, valves, and tallow pots were intact, albeit heavily camouflaged. Again this was totally unexpected, for every other steamer in shallow water close to a centre of population like Port Gregory had long since been stripped of all its brasses and copperwork by salvors and recreational divers.

As a result, it was decided that, after the removal of loose sedentary colonizing marine life (weed, sponges and seasquirts, but not the concretion layer) the engine would be subject to very detailed recording and measurement.

4.5 Site survey and test excavation

While the three-day predisturbance survey was under way, recording of the machinery and the production of a two-dimensional and three-dimensional photographic record were commenced. The air supply chosen for was hookah, or compressor-driven surface supply type, which in the shallow waters on the *Xantho* allowed unlimited dive time. This also allowed the engine to be recorded in considerable detail (see below).

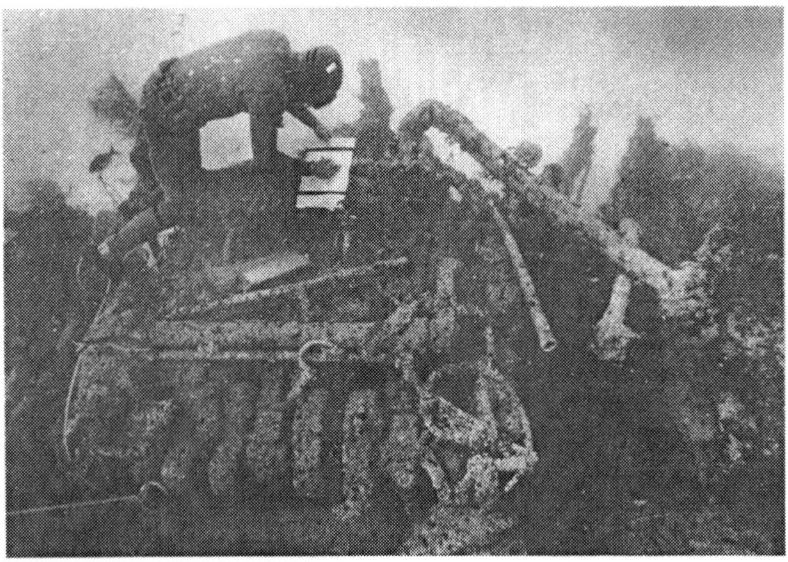

Figure 30. Geoff Kimpton recording the *Xantho* engine using manual methods. The crankshaft, oil cups, and piping are visible (M. McCarthy).

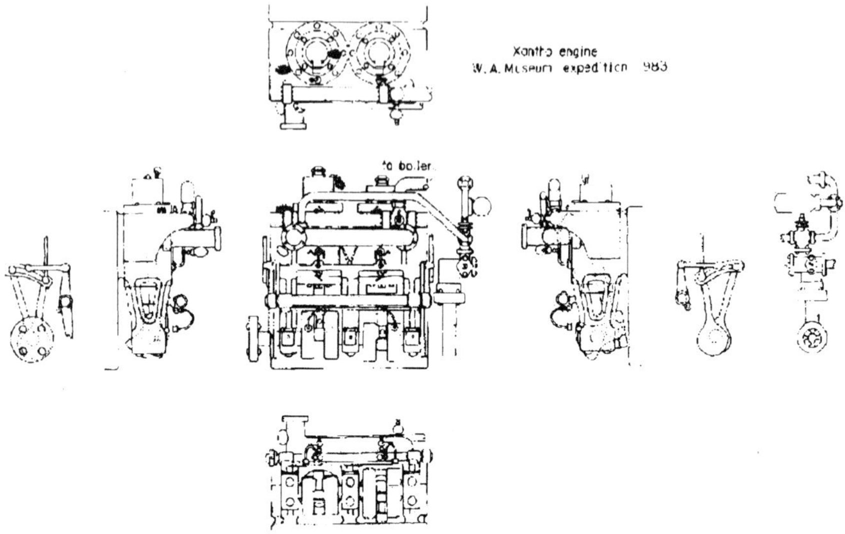

Figure 31. Drawings of the SS *Xantho* engine in its concreted state (Geoff Kimpton).

Current was by far the most trying factor, making normal anchoring sometimes impossible and (contrary to earlier expectations about the benign nature of the site) rendering diving quite dangerous. Days of strong current were invariably days of moderate to poor visibility with large banks of loose weed moving throughout the site, especially on change of tide. The weed reduced visibility and collected on grid wires, tapes, and even on the diver's gear. On occasions a cloud of weed would suddenly descend upon the wreck, reducing visibility to less than 1 meter within only a few minutes. Here was one reason why sports divers had not had an opportunity to properly examine the wreck and then begin the recovery of the copper and brasswork!

Because the hull appeared disarticulated or at best twisted around its longitudinal axis it was examined at a number of locations using builders' levels and steel carpenters' tapes. It was subsequently found to be broken in three places. Under the engine the hull dropped 3° toward the stern and inclined 8° to starboard. Aft of the engine it dropped 11° toward the stern. In contrast with the situation aft, fore of the engine the boiler dropped 8–9° toward the bow and was inclined 6° to starboard. Measurement of a wooden deck stanchion (or mast) forward of the boiler indicated a lean to starboard of 14° in the fore part of the ship, i.e., *Xantho* lay in four discrete sections, each with a different alignment.

Though broken into four sections, each has moved only centimeters apart, thus remaining in close proximity to other sections. All but a few unsupported parts of the hull had collapsed downhill to starboard in the direction of the prevailing current, and not to port in the direction of the seas and swell as originally expected. Finally, it was concluded that the breaks appeared to have occurred as a result of the movement of sand under and around the hull and that the resultant forces caused the propeller shaft to break away from the engine at its coupling with the thrust block (see Figure 54).

4.6 Test excavation method

In respect of the original artefact management brief, the sampling method employed was a visual examination of the surface remains and then an examination of deposits by conducting a series of test trenches. These were excavated along the length of the ship outside the hull, across the site at the bow and stern, and then through the hull at 4-meter intervals. This particular sampling technique is cognizant of both the disintegration processes that result in the spread of artefactual material outside of the hull as a wreck collapses or opens up, and of the traditional compartmentalization of shipboard activity.

On iron sailing ships and early steamers, officers and passengers were usually housed aft in the poop and the crew forward in the forecastle. This also appears to be the case with the *Xantho*. (See, for example, evidence at the court of inquiry into the loss of *Xantho* in Chapter 3). The quarters of the officers and passengers in a large ship are also usually separate compartments. The separation is an almost rigid feature of shipboard life based on centuries of European seafaring tradition. The boundaries are usually crossed where officers or passengers require service, where work is to be performed, or by invitation to functions. On smaller ships, officers and passengers would dine together. Even more rigid conventions are seen on large naval vessels where entire classes of sailors ate and lived in separate compartments and where a ship's master could have completely separate accommodation, eating alone and served by a personal steward.

Accommodation, machinery, and cargo spaces themselves are also discrete, specific-purpose compartments within a ship. These are separated by barriers deliberately designed to minimize unauthorized or unwanted access by people or materials, especially noxious engine wastes and seawater.

Though the movement of water through a wreck and the collapse of decks down onto lower ones can markedly affect compartmentalisation, the artefact distribution and the structural remains can still reflect specific uses or classes of activities within shipboard life. With iron and steamship wrecks, especially where the remains of these compartments are visible above the seabed, an opportunity emerges to structure research design such that it enables the inquirer to target specific areas while leaving others untouched. Where structural elements such as bulkheads, decks, or other features still exist, an excavation of one discrete area need not necessarily impinge on another. Further, while there are numerous examples where maritime archaeologists (e.g., Blawatsky, 1972; Ruoff, 1972; Stevens, 1982; and Ringer, 1983) have identified layers in underwater deposits, it is often a very difficult undertaking due to the downward movement of sediment during the excavation itself. *Coning* or the production of an uncontrollable funnel-like test-pit is often the result. The compartmentalization of a wreck into its various social and functional units and the many flat surfaces throughout can assist in overcoming the difficulty experienced in maintaining a vertical face in underwater archaeology.

The excavation tool used on *Xantho* was a water dredge, an excavating system common to underwater archaeological work (e.g., Green, 1990). In this particular phase, overburden was removed from defined areas within a 1-meter-wide trench along the survey lines delineating the site (Figure 32). Area excavation, the norm at colonial-period wrecks in Western Australia at the time (cf. McCarthy, 1998a), was not considered feasible or desirable.

Where excavation was conducted through the layer of mobile sand and loose clay on the *Xantho*, the customary backfilling of trenches after excavation and recording was not necessary due to the rapid ingress of sand. This was of such a speed that within a few hours after excavation the trenches were again filled. All visible indications that an excavation had taken place disappeared overnight, such was the mobility of the sea-bed in the prevailing current.

4.7 Results of the survey and test excavation.

The drawing, measurement, and inspection of the hull, boiler, engine, driveshaft, windlass, deck winch, and other machinery were successfully completed. Due to their relative size and dominant nature, recording the engine, boiler, stern, and stem of the *Xantho* proved problematic, however. Eventually a combination of manual recording and photographic methods was applied to the problem and a site plan was produced by trilateration and by right-angle offsets (Figure 32). Common two-camera or stereo overlap photogrammetric techniques or manual three-dimensional methods also proved suitable. This resulted in a plan of the wreck, a plan view photomosaic of the stern section, a port elevation photomosaic of the entire site, a manually recorded three-dimensional drawing of the engine and other machinery, and two-dimensional and three-dimensional photography throughout, including the engine.

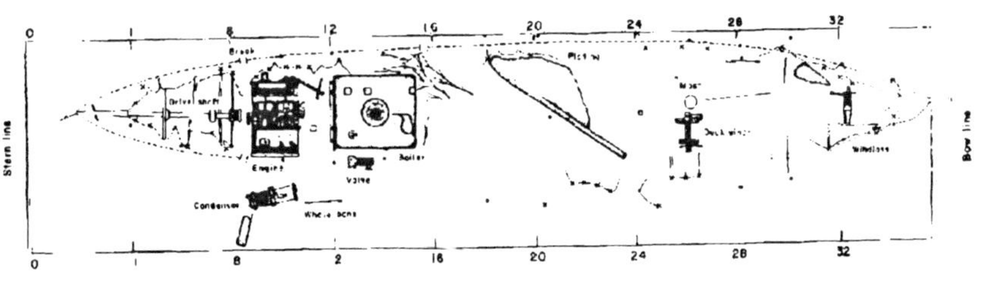

Figure 32: SS *Xantho*, site plan (Stephen Cushnahan).

The test excavation indicated that the wreck lay on a 1-meter thick mobile seafloor of soft sand over a thin band of hard clay, which in turn overlaid a thick weed mat on what was believed to be a sand bottom. The weed mat proved almost impossible to penetrate. The plan view of the site shows the wreck has opened out downcurrent and this was highlighted by the spread of lead ore and the presence of machinery outside the hull to starboard. The ore provides a near-impenetrable protective mass for material and structure lying below. The presence of the thick weed mat below the mobile layer of sand leads to the conclusion that the wreck forms a barrier to the movement of weed in the current. Consequently, weed would have quickly filled the spaces inside the vessel and under the hull. Mobile sand would have helped compact this deposited weed and, as a result, preserved material could be expected to lie buried beneath it.

No loose artefacts were seen on the seabed. Excavations along the perimeter of the wreck on both sides and through the bow and stern compartments also revealed few artefacts. Only the area immediately aft of the boiler appeared fertile. In total, 35 items (XA 7–XA 41) were located and raised, to add to the six artefacts raised on the 1979 inspection. In general, they were categorized as material common to nineteenth-century European seaborne life, such as ship's fittings, cargo items, personal items (such as the sole of a shoe), and glass and ceramics from the galley or cargo. At variance from items usually recovered from late nineteenth-century wooden hulled wrecks were iron hull fragments.

Evidence of site contamination was noted, though it was located deep in the mobile sand layer. This was in the form of modern material, such as light globe fragments and a motor vehicle oil filter. While the oil filter was immediately discounted, the presence of the light globe required some analysis. Research indicated that electric light was yet to become a feature of such vessels and oil-burning lamps were used instead. The saloon of the Inman liner *City of Berlin* was lit by electricity in 1879, for example, and it represents one of the first instances of the use of that technology (Smith, 1937). A coconut husk and whalebone were also possibly contamination from Port Gregory's early days as a whaling station (Trenaman, 1934; Heppingstone, n.d.), though they could have constituted galley supplies and cargo from Broadhurst's activities further north. The lead ore and some of the bagging in which it was stowed was still visible in the cargo hold forward of the boiler, and samples of both were recovered. Rough-cut branches were found throughout the cargo spaces, possibly representing "dunnage" (or softeners on which cargo was laid to prevent damage to the vessel's hull). The use of dunnage was mentioned in the court of inquiry into the loss of the ship.

No evidence was found of the compartmentalization of the ship into two accommodation areas—for the forecastle was not closely examined partly due to the weed mat and also due to the focus on Broadhurst, who would have occupied quarters in the stern.

Sections of the hull had considerable relief, especially the bow and stern sections. Much of it had collapsed, however, and the port side of the vessel appears to have collapsed into the ship. In contrast, with the exception of the hull near the stern and a section just forward of the boiler, the entire starboard side has collapsed outward and downcurrent (see Figure 26).

The inner and aft walls of a coal bunker were located on the starboard side of the boiler. Forward of the boiler was the support for the starboard sponson that once held the bearing of the paddle wheel. The sponson also served to transmit the thrust of the paddle to the hull and needed to be stronger and thicker than the surrounding frames and hull plates. As indicated earlier, even this strengthened section of the hull was fragile and corrosion appeared to be continuing within the concretions.

It was estimated that the hull was buried to a depth of a least 1.5 to 2 meters forward and slightly less aft. It soon became evident that, of the visible remains, the boiler and engine room machinery were the most intact part of the vessel.

As indicated, the engine was identified as a small, simple expansion, horizontal trunk engine. The 16-cm-diameter propeller shaft was quite short, extending only a distance of 5.5 m. Though difficult to confirm due to concretion, what appeared to be a "thrust block" and a "dog clutch" (or disconnecting device) appeared on the shaft. The shaft itself was supported in three places—on a thrust block bearer, on one "stool" or "plummer block" and by the stern tube itself (see Figure 45). A detailed examination of the shaft, bearings, and stern tube was not made due to the concretion layer. The thrust block appeared similar to a common small thrust block with wick lubricators to each collar positioned on an open oil box, which was mounted on top of the thrust block, however (Jamieson, 1897).

The propeller was an iron screw of approximately 1.8 meters (6 feet) diameter, situated inside a stern aperture forward of the sternpost. Only one vertical blade was visible, though this was enough to indicate that it was "right-handed" i.e., the propeller (and hence the engine) rotated clockwise when viewed from astern. Detailed measurements of size and pitch were prevented by the sand buildup and strong currents which prevented accurate usage of a plumb bob.

The single-ended, two-furnaced, return-tube boiler measured 3.2 meters (10.5 feet) in length by 2.2 meters (7.2 feet) in diameter. It appeared to have a slightly elliptical shape and both the furnace doors were shut, precluding an inspection of the interior. A number of concreted brass fittings were noted on the forward face. These were left undisturbed. A large unidentified valve, probably a relief valve that appeared to have been knocked off the boiler, was noted lying on the seabed on the starboard side.

Also lying to starboard of the engine, and apparently having fallen from a position on the starboard hull in the engine room, was what appeared to be a condenser (Figure 32). As indicated in the section on marine engineering earlier, condensers were used to recycle expended steam and one was expected on the *Xantho*. Its form in this instance was a puzzle, however, and no driving rods connecting it to the engine were found.

4.8 The application of anodes

As indicated, it had become apparent that the ship was of considerable regional importance. The engine was an uncommon type, one normally found in a naval context and its presence on the *Xantho* required some explanation. It was intact, one of the only complete nineteenth-century engines left in the waters of Western Australia, but it was also physically degrading. Alerted by the museum presence, divers and the fishing community were showing a renewed interest in what they had previously thought to be a worthless wreck and it was feared they would attempt to recover the brasses and the copper as soon as the expedition left.

The site had not yet been declared historic and the options for the preservation of the engine in the face of the legal vacuum were broadly canvassed. It was eventually decided that after the recording of the engine and other features was complete, a covering of rocks would serve to provide camouflage and would also serve to deter idle looters.

In examining the engine more closely, it also became evident that a large, four-bolt flange on the aft section of the crankshaft had parted from its partner, which was originally connected to the propeller shaft forward of the thrust block. This apparently occurred as the hull of the *Xantho* broke up, leaving the engine disconnected from the shaft and thrust block (see Figure 53). This also left the engine electrically isolated from the remains of the hull, stern, and propeller shaft. As a result, the possibility of applying anodes to the engine was also mooted.

It was hoped that this conservation technique would reduce the toxicity of the copper and allow a relatively rapid secondary colonization by marine organisms (North to McCarthy, *Expedition Daybook*, May 9, 1983). This regrowth was considered a priority in order to further camouflage the engine.

The anodes would also begin the process of preservation of the underlying metal and would prolong the life of the engine on the seabed. The technique was standard practice in underwater environment, where it was utilized on working vessels, steel jetty piles, oil rigs, and the like. Its application to a shipwreck had still to be tested, though it had been mooted for use at the wreck of the USS *Monitor* (National Trust for Historic Preservation in the United States, 1978).

North's proposal to experiment with anodes was adopted after some discussion, and he proceeded down to his laboratory at Fremantle in order to prepare them for shipping back to the site. In his instructions for the application of the anodes North wrote:

> The engine of the *Xantho* is in surprisingly good condition considering its age, the presence of many galvanic couples, and the underwater environment. The attachment of sacrificial anodes, in May 1983, will prevent any further decay and should actually start in the preservation treatment by encouraging the release of chloride salts from the corrosion products (N. North, personal communication, May 1983).

The anodes were subsequently attached to a counterweight on the crankshaft and to the propeller shaft aft of the thrust block. A circuit was made by winding down a pointed screw, held in a bracelet (a converted clamp), through the concretion to the original metal.

After the anodes were attached to the engine, it was covered with rocks obtained from a nearby creekbed. All grid wires, and other equipment, markers at each extremity of the site, were removed and the site was closed.

In the process of this first study, 200 operator-hours had been spent underwater over a total of 76 operator-days worked on the 10-day expedition (including travel).

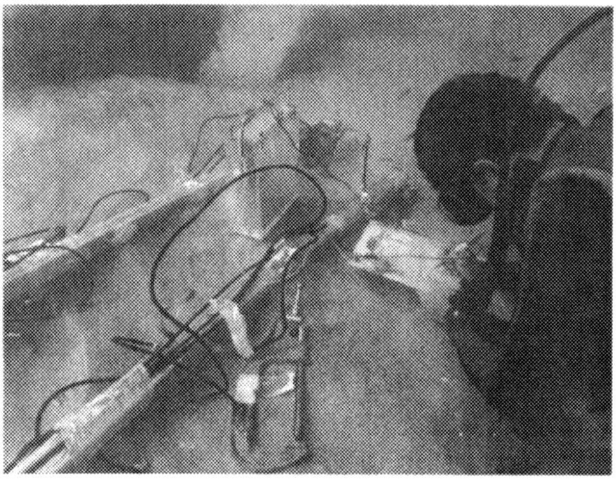

Figure 33a. The anodes before application to the wreck (Jon Carpenter).

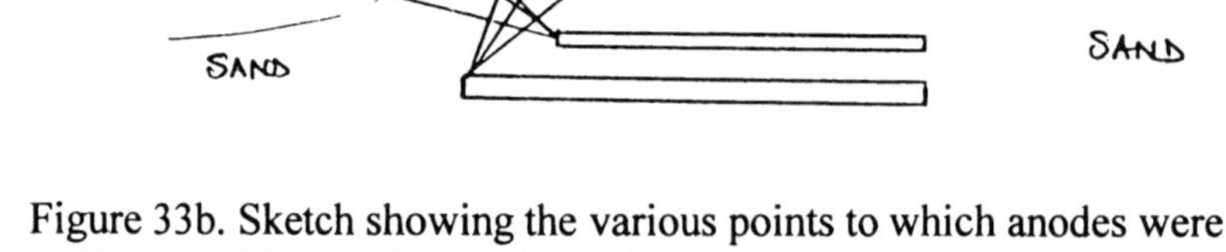

Figure 33b. Sketch showing the various points to which anodes were fixed on this occasion and on subsequent dives (Jon Carpenter).

Chapter 5

Site formation processes

Controlling for the variables that formed the Xantho site

This chapter aims to "control" or properly account for postdepositional processes at the site—a standard practice in archaeology generally. This requires us to examine iron and steamship wrecks as a *class* or suite of archaeological sites and to proceed from there in a comparative fashion to the *Xantho*. These are necessary steps, for as Gould (1990:48) has noted:

> In looking at general relationships between behaviour and material residues, the first thing to consider is the total ecosystem in which this behaviour takes place.

Attempts to control for postdepositional processes in maritime archaeology represent an extension of work by Muckelroy, who attempted to identify features common to all shipwrecks, including disintegration processes. He noted that if "regular features common to all instances" can be identified, then their implications for the analysis of the material remains at shipwrecks can be ascertained. He also indicated that the "validity of any conclusions reached" was dependent on the understanding of these processes (1978:157).

5. POSTDEPOSITIONAL PROCESSES AT AN IRON WRECK

That sections of nineteenth-century iron or steel hulls can be found above the sea floor in high-energy aerobic waters represents a departure from the usual situation at nineteenth-century wooden-hulled sites. Though occasionally exposed due to seabed movement, the remains of a wooden hull are normally found buried beneath ballast, sediments, cargo, or similar protective materials.

Though iron or steel hulls are often found as a semi-intact unit, even in shallow waters, like their wooden counterparts they can be destroyed by wave action in a very short time. The seas can also move entire parts of an iron or steel ship, including nonbuoyant sections of the hull, considerable distances from the parent wreck. Elements such as the nature and topography of the seabed, the depth, the "fetch" of the seas, the type of hull, its integrity at the moment of sinking, the type and weight of cargo and/or ballast carried, and many other factors are of relevance in that process. Muckelroy identified some of these phenomena in general terms as "scrambling devices" or factors that served to rearrange the elements of the vessel and to alter the ship after it is wrecked (1978:167).

5.1 The effect of corrosion and concretion

With reference to corrosion (a major scrambling effect, in that it weakens a structure over time), Muckelroy noted that it can vary across quite short distances on a site, depending on seabed type, marine growths, the presence of dissimilar metals and other chemical and electrical phenomena. It is also affected, and often masked by, the presence of the substance mentioned earlier called *concretion*. This is itself a complex phenomenon, appearing on most metallic surfaces underwater, especially where the water is comparatively warm and the wreck and its contents lie in a predominantly aerobic environment.

In the first detailed analysis of the phenomenon, Neil North found that concretions were formed on stationary and biologically nontoxic material, such as iron, within the first few months of wrecking. Initially this was due to coralline algae, which, unlike soft algae, has a partial exoskeleton of calcium carbonate ($CaCO_3$). As coralline algae die, their exoskeletons remain and are subsequently overlaid by later growths of the same materials. These buildups merge progressively with adjacent objects, often forming a large mass on which a secondary growth of seaweed, soft corals, molluscs, and other life occurs. The rough outer surface of the concretion then provides a trap for sand particles, coral fragments, and other debris being moved around the site. Material can become completely covered with a thin layer of coralline algae and secondary growths within about 12 months.

It was also found that though the concretions found on iron were externally indistinguishable from those formed on natural materials, they formed a layer of low porosity on the surface of the iron. This retarded the movement of corrosion products away from the rusting iron and resulted in the production of an acidic iron-rich corrosive solution. These chloride concentrations can rise by a factor of three and the pH inside the concretion can drop from around 8.2 to 4.2 for example, thereby increasing acidity considerably (North, 1976). North's research showed that corrosion continues within layers of concretion, though it is most often at a reduced rate. It can do so until the last of the oxygen bound in the form H_2O is consumed and hydrogen is given off. This can leave nothing of the original iron after a few hundred years. Thus while it is acknowledged that concretion can serve to "protect" the iron beneath (e.g., Oddy, 1975:367), corrosion continues and it can do so until all the iron is consumed.

As indicated, North and his team showed that on the *Xantho* corrosion was not proceeding at a uniform rate and was continuing even in an anaerobic environment under the concretions. Apparently intact, heavily concreted iron structures, such as the windlass and deck winch, looked strong, but were shown to be hollow shells. This effect has been long-recognized in maritime archaeology (e.g., Van Doorninck, 1972).

The corrosion specialists also found iron and copper sulphides in some of the concretion matrices examined at the *Xantho* site. These indicated biological action by sulphate-reducing bacteria. In finding evidence for the bacteria at work during some of the excavation phases, they were able to conclude that corrosion was also continuing in the sediments in which the vessel was buried. While they did not pursue this avenue of inquiry, this is an important issue, for it was once believed that iron hulls buried in sediment would be relatively well preserved. This is not to imply that some degree of protection might not occur as a result of sedimentation processes. Rather it is to acknowledge that an iron or steel hull lying buried in anaerobic seabed sediments is still actively degrading. This process is defined as "the deterioration of metal by corrosion processes which occur, either directly or indirectly, as a result of the metabolic activity of microorganisms" (Evans, 1973:469). These effects were discovered over half a century ago (e.g., Uhlig, 1948) and they have long since been noted on shipwrecks (e.g., Arrhenius *et al.*, 1973; Barkman, 1977).

These particular bacteria can be categorized as either aerobic or anaerobic microorganisms, depending on their viability in relatively high or virtually zero oxygen levels. Both types require organic and inorganic chemical compounds from which to obtain oxygen, carbon, nitrogen, hydrogen, or sulphur. Other factors such as pH, oxygen concentration and temperature are also crucial to their growth.

The implications of the presence of sulphate-reducing bacteria for the survival of iron are profound. It was shown a half-century ago, for example, that even noncorrosive, washed, silica sands can be rendered corrosive beneath an apparently sterile seabed (cf. Hadley, 1948).

In examining the end result of corrosion and other transformational factors, Muckelroy provided a five-stage site-classificatory index (1978:164–165). This was based on the degree of hull and material survival, ranging from his Class 1 sites, with extensive structural, organic and other remains in a coherent distribution, down to Class 5 sites. These have no defined structure and the few material remains lie scattered in an apparently disordered fashion over the seabed.

5.2 Muckelroy's index applied to modern sites

In utilizing this classificatory index, iron and steel wrecks could also range down from Class 1 to Class 5 archaeological sites. Arguably a completely intact and formerly fully operational submarine could fall into Class 1. An example is IJN *I 124* (1926–1942), a Japanese submarine lying virtually untouched with its entire crew and their effects near Darwin in northern Australia (McCarthy, 1998b). The historic American Civil War submarine H.L. *Hunley* (1863–1864) which lies almost totally buried with its crew off Charleston in South Carolina (Murphy, 1998), is another example.

There are numerous other examples and unlike surface vessels most form a virtually unbroken "capsule" in the true sense of the word. This feature provides both the strength and weakness of the submarine as an archaeological site, for they are also the most easily removed from the seabed and can become an easy target for well-financed adventurers, museologists and entrepreneurs. One example is the German *U 534* (1942–1945) that was recently raised from Danish waters and then placed on exhibit in Britain (see the discussion in McCarthy, 1998b).

The early British submarines *Resurgam* (1879–1880) and the *Holland 1* (1901–1913) were lost while under tow, and in not being in an operational mode would form a subgroup to the highest wreck site classification, partly due to their historic status. A very low, but nevertheless significant, category would be afforded the six WWI *J boats* stripped and scuttled off Melbourne in the 1920s. These submarines represent a unique opportunity to examine the corrosion status of a series of almost identical vessels, built in the same yard within years of each other, with similar service records and similar disposal dates. The chief variables appear to be depth and site conditions; i.e., they represent a class of deliberately disposed sites still possessing technological value to researchers in history, corrosion science, and site formation processes (McCarthy, 1998b).

The steel-hulled RMS *Titanic* (1912–1912) (e.g., Ballard, 1987) and the USS *Monitor* mentioned previously could be in a similar class to a relatively untouched fully operational and enclosed historic submarine. Their status is slowly being eroded, however. Until recently, they were inaccessible, relatively intact hulls that had not been subject to *primary salvage* (the recovery of materials by their owners, operators, or agents). Sites similar, yet subjected to primary salvage, are the USS *Utah* (1909–1942) and the USS *Arizona* at Pearl Harbour (Lenihan, 1989; Lenihan and Murphy, 1989). *Arizona* is now inaccessible to divers by legislation and has not been affected by *secondary salvage*, or the action of professional salvors or sports divers in recent times. *Titanic* is an interesting contrast in that, while its hull was being left relatively undisturbed, its debris field was being harvested by commercial and museological interests; i.e., secondary salvage was occurring at the site until recently. In comparison to USS *Arizona*, which was undergoing only natural change, the class to which one could allocate *Titanic* was gradually slipping.

The SS *Yongala* (1903–1911) on the Australian Great Barrier Reef is one of the most visited and regularly monitored of all iron and steamship sites (e.g., Riley, 1994). Despite the high visitation rates, considerable amounts of loose artefactual material remain to be viewed by the diving public in what has been referred to elsewhere as an "underwater display case" mode (McCarthy, 1981). Skeletal material, and numerous fittings, fixtures, and artefacts were visible in 1994, for example. Though now well managed, *Yongala* was subject to uncontrolled postdepositional transformations in the past and would lie in a subclass within Muckelroy's highest category (e.g., Class 1b-c).

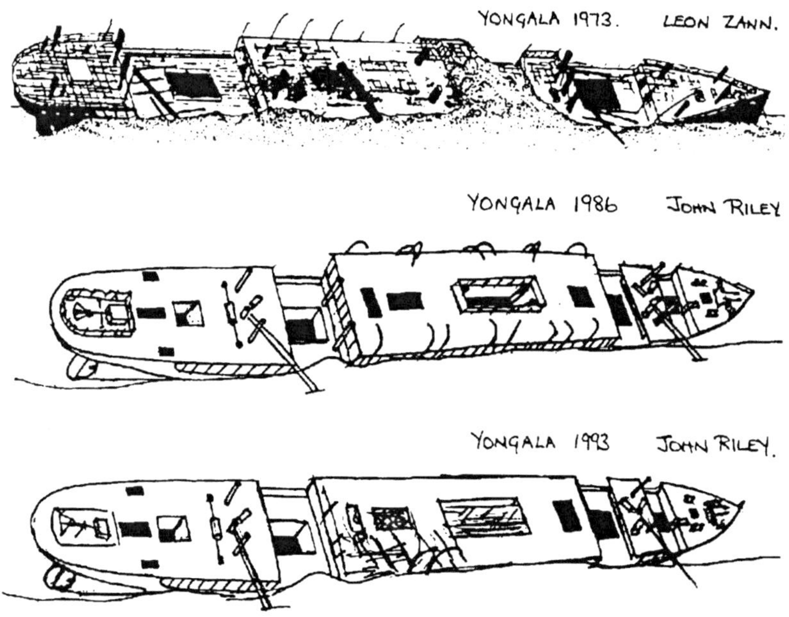

Figure 34a and b. The SS *Yongala* (Leon Zann and John Riley). This wreck was once also being degraded by uncontrolled anchoring, resulting in the loss of fragile stanchions and davits. Note also the seabed changes.

The intact wreck of the SS *Sunbeam* (1861–1892), lying off the remote Kimberley coast of Western Australia, would fall into a lower class entirely. While the bilge is expected to be a rich source of materials, there are few organic materials visible and little bar the hull and heavy machinery remains above the seafloor (Sledge, 1978; Henderson and Sledge, 1984).

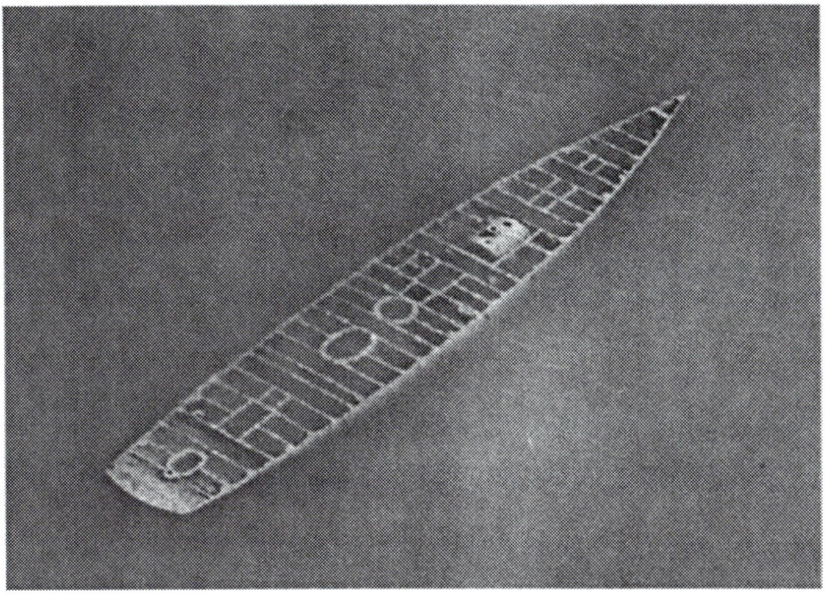

Figure 35. The SS *Sunbeam,* showing an apparently intact hull minus its wooden decking (R. Coulter, Australian Customs Service).

A lesser category again could be given to the remains of the SS *Macedon* (1870–1883). It was subject to heavy primary salvage and to continued secondary salvage with the onset of sports diving. In recent years it has become a part of a popular "wreck trail" concept and has thereby been returned to the "systemic" or living context (McCarthy, 1983). A comparison of the two illustrations below (the photograph is from the early 1980s and the isometric from the early 1990s) shows that the wreck is still slowly deteriorating.

It is not intended to dwell on Muckelroy's classificatory scheme other than to note the possibilities for further study and that it is not a static indicator. It should also prove of use in helping assist in formalizing descriptive and analytical studies, such as the following observations about iron wreck burial and site disintegration.

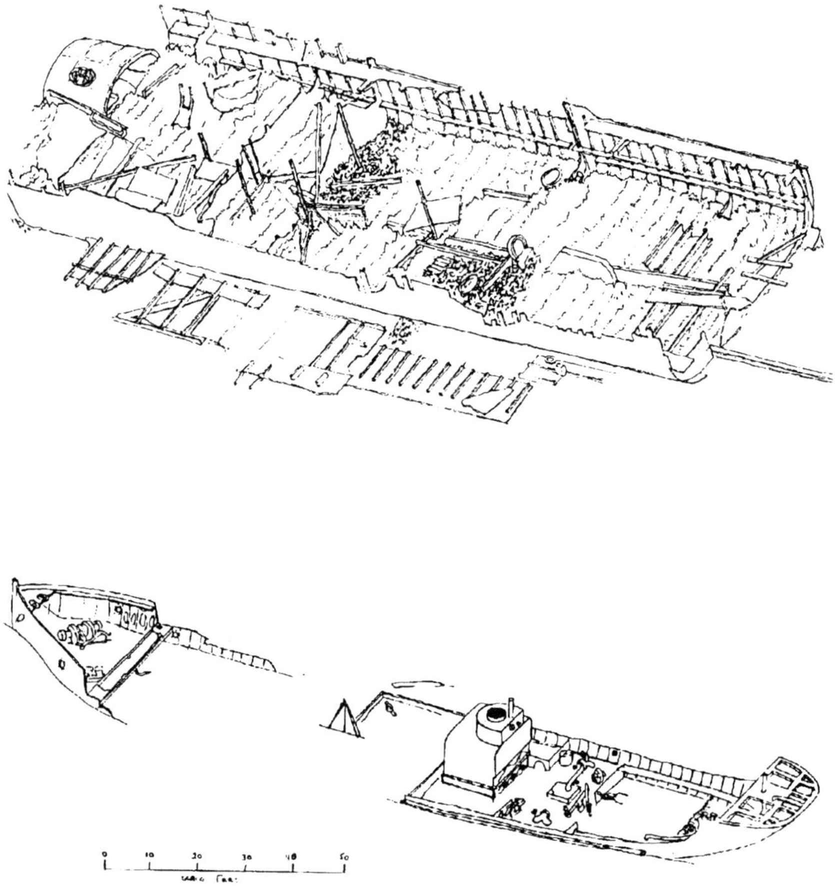

Figure 36a and b. The SS *Macedon*. Over the space of these two illustrations the deck beams have totally collapsed, leaving the sides of the vessel unsupported (photograph by Pat Baker, isometric projection by Colin Cockram).

5.3 John Riley's observations : The waterline theory

The extent of burial of a wreck in sediment or other matrix is a major factor in physically limiting the movement of fixtures, fittings, and artefactual material from the hull. Observations about the commonalities of iron hull burial were first made by Australian avocational practitioner John Riley following a decade of study of over 100 wrecks off the coast of New South Wales (e.g., Riley, 1988a). He concluded that ships generally sink to the waterline when they come to lie upright on a seafloor of sand or similar soft sediment and he came to use this observation to effect as a predictive tool in searching for other sites (see illustrations in Figure 9).

Riley's many wreck reports (most are in deep water), the use of isometric projections, cardboard site models, and (with Mike Lorimer), the description of sediment layers within the hull of *John Penn* (1867–1869) were important developments in the field of iron and steamship archaeology (John Penn Project Team, 1984; Lorimer, 1988; Riley, 1988a).

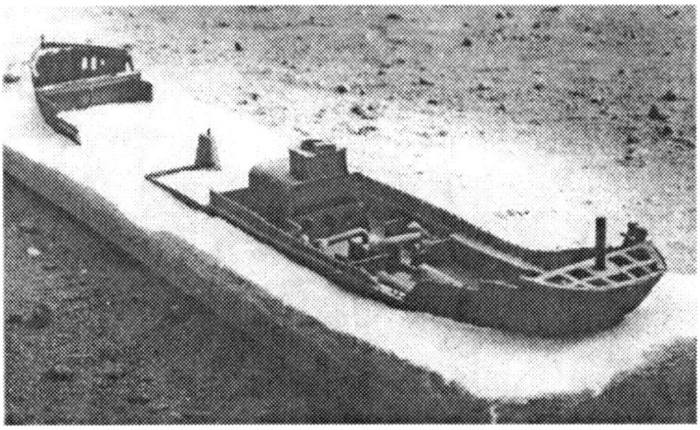

Figure 37. The remains of the TSS *John Penn* (John Riley).

Riley also noted that boilers eventually roll out of steamships lying in the surf zone or in heavy conditions. Often they come to rest on their end where they present least resistance to seas and swell. These patterns are illustrated in Figures 9 and 38. In the case of the SS *Windsor* (1890–1908), on the Abrolhos Islands, the gradual sinking of an upright boiler into an apparently hard reef platform was observed.

a

b

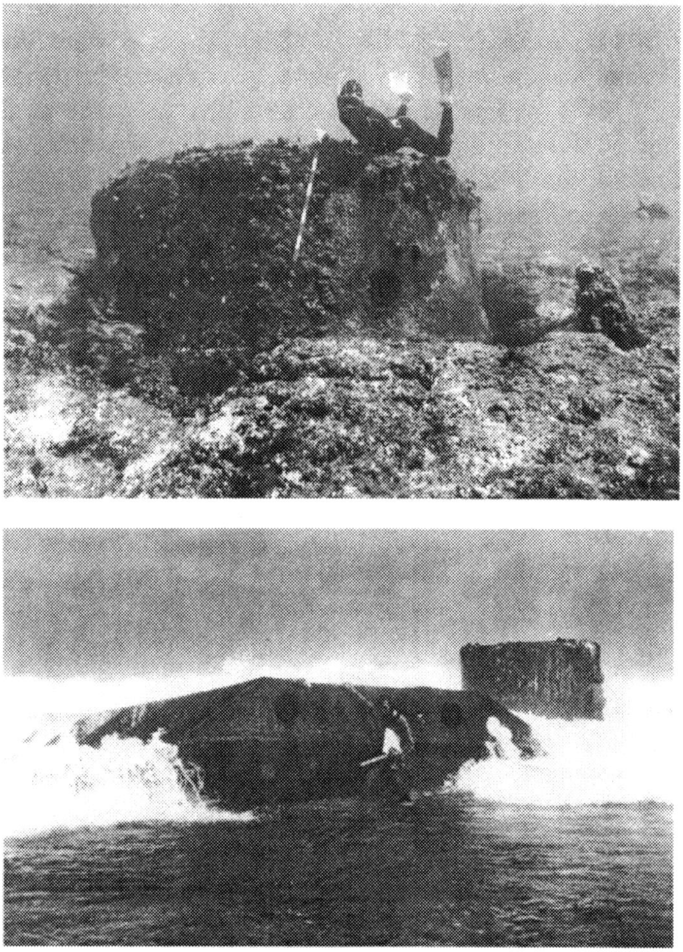

Figure 38a and b. A *Windsor* boiler on the seabed and another alongside an upturned section of the vessel's floor on the reef platform. The underwater scale is 1 meter long (M. McCarthy).

At *Windsor* there are two identical boilers on-site and both lie on their end. One has sunk into the submerged reef alongside the engine, where it is exposed to the constant pulsating action of seas and swell. The other lies totally exposed on the drying reef platform adjacent the wreck. In contrast to its "pair" on the seabed it has not dug its way into the reef. When consulted, a diving geologist advised that this had occurred because it lies on a harder platform and because it receives unidirectional seas that are normally broken by the reef before they impact on the structure (R. Brown, personal communication, November, 1979)

5.4 Other commonalities observed

It is also evident that when lying upright on a hard unyielding sea-bottom, the midships section of an iron-hulled vessel tends to break at the turn of the bilge. The hull then lies almost flat on the seabed and materials lie either side, or on the hull, depending on the direction of the prevailing seas, e.g., the barque *Ann Millicent* (?–1890), which dries at low-water spring tides on Ashmore Reef midway between Australia and Indonesia.

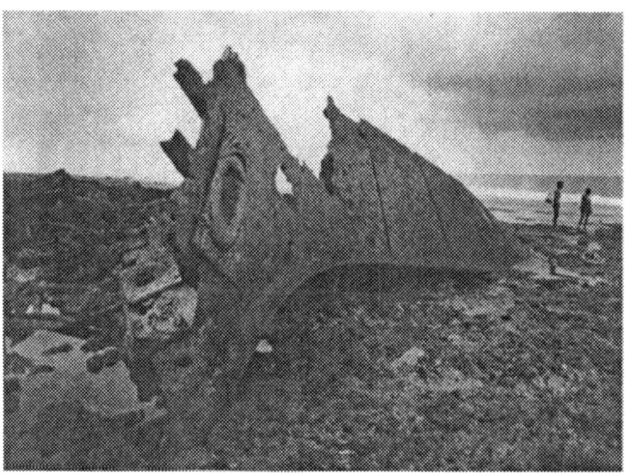

Figure 39. Indonesian divers at the bow of the iron barque *Ann Millicent* (Jon Carpenter). Though lost in 1890, its plates are fixed in clinker style, indicating that it might have been built much earlier.

In the case of both steamers and sailing vessels, the bow and stern "triangles" (so-called due to their characteristic shape) can remain standing for quite a long time. Then they eventually break from the keel/keelson, and in falling over, lie on the seabed as shown in the illustration above and in the *Denton Holme* (1863–1890) and other cases following. Their relative longevity, in comparison with cargo and midships sections, is due to their immense strength and cross bracing.

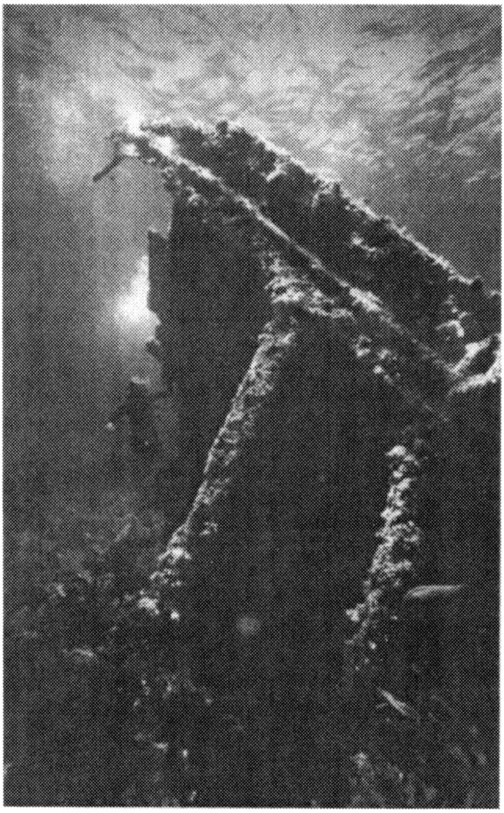

Figure 40. The bow "triangle" of the iron barque *Denton Holme* (Pat Baker).

The stern of the iron barque *Ben Ledi* (1868–1879) in shallow water on the Abrolhos Islands in Western Australia, is another example of this phenomenon.

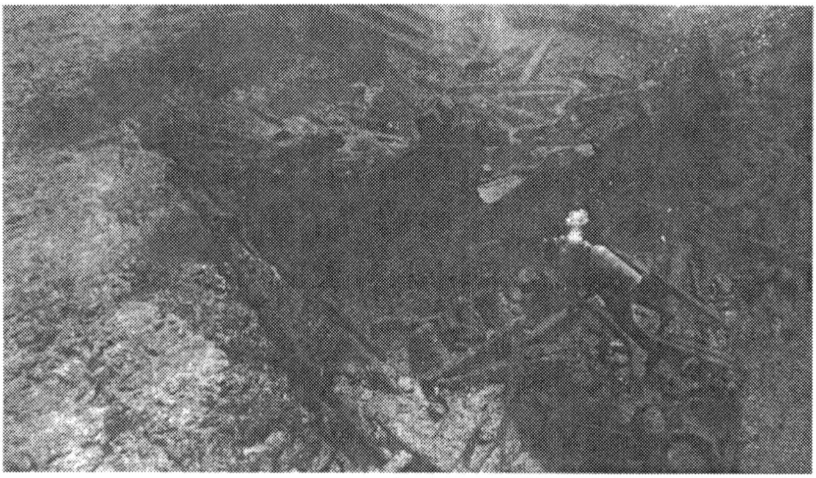

Figure 41. The stern of the *Ben Ledi*. The hull is visible in the background. (Pat Baker).

Eventually an exposed iron or steel vessel assumes a characteristic appearance; i.e., bow and stern triangles lying on their side, separated by flattened cargo holds, and (if they are present) a machinery section dominated by engines and boilers.

The wreck of the iron-hulled *SS Colac* (1895–1910) in King Sound near Derby (Figure 3 and 42 below) displays some of the patterns noted above, such as disjointed but otherwise intact bow and stern triangles and collapsed cargo sections. After being damaged and rendered unseaworthy in 1910, this vessel was run aground on a sandbar and abandoned after heavy primary salvage.

Though subject to diurnal tides, sometimes in excess of 10 meters, with extremely strong currents, the wreck lies deep within a bay and does not experience heavy swells (RAN, *Australian Tide Tables*). The ship has sunk to near the waterline in its own scour pit and the bow and stern triangles and cargo holds are as expected. There is an apparent anomaly, a completely intact midships section, however. On inspection, it became apparent that the relative longevity of this section is due to the longitudinal and lateral strength of the combined engine, bunker, and boiler rooms.

a

b

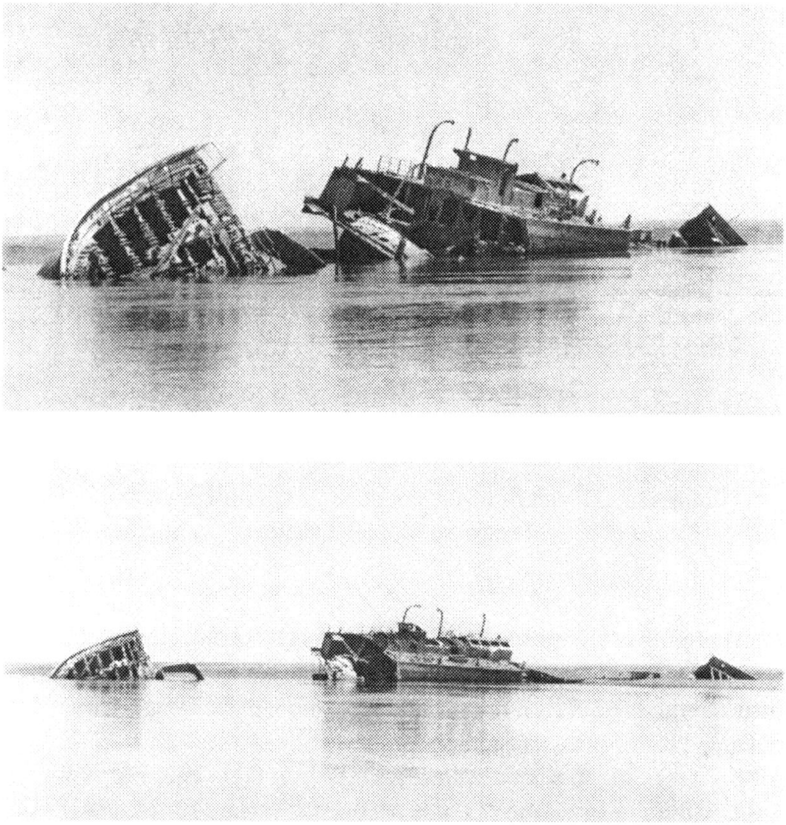

Figure 42 a and b. Views of the SS *Colac* taken in October 1995 at low water spring tides near Derby on the Western Australian coast (M. McCarthy).

These compartments form a very strong unit, which is heavily braced both horizontally and vertically in traditional fashion and then by a heavily built coal chute at an angle of around 45° from the vertical in the bunker room. The prevailing site conditions of strong, but directionally constant, tidal currents and the absence of the constant pulsing action of seas and swell represent other significant factors. Further, biologists are yet to visit the site and to provide comment on the intertidal shellfish and other life, which may be important elements in that they protect the hull from the effect of wind and wave.

This contrasts with the former WWII Liberty Ship SS *Alkimos* (1943–1963) a welded steel-hulled vessel that lies in open waters subject to swell, but with a tidal range of only 1–1.5m. Though lost nearly half a century after *Colac*, when photographed in mid-1990, only 30 years after it was abandoned, little remained of the *Alkimos* hull in what is effectively a constant 1–2 m high "splash zone" between wind and wave (Figure 43a). At *Alkimos,* highly oxygenated saline water at the sea surface constantly impinged on this small intertidal zone with well-recognized effect (cf. La Que, 1975). This is in stark contrast to *Colac*, which is subject to 10 meter tides. Being in a Mediterranean climate, in comparison to *Colac*'s tropical location, the colonization by marine animals and growth is markedly different at these two sites, requiring some attention by the appropriate specialists.

An interesting example of the need to control for postdepositional effects is again found at *Alkimos*. The apparent total collapse of the wreck forward of amidships and an intact bow triangle (Figure 43a) are as expected of any natural collapse. In this case these features are actually the result of documented cultural transformations in the form of secondary salvage over a number of years by scrap-metal merchants (Nairn and Sue, 1975). This instance highlights the need to account for all postdepositional factors before making comment on the processes involved.

Alkimos also allows us to demonstrate the fact that the process of site disintegration at steel wrecks can be quite dramatic. When the wreck was examined in April 1999, aft of the bow only the hull surrounding the engine room was left standing and it too was collapsing into the sea on the starboard, or weather, side (Figure 43b). It was evident that in a short time all that would be left above the sea surface will be the bow triangle, the engine and some other heavy machinery, such as the condenser and boiler. In time, the bow triangle will collapse onto its side, leaving only the engine, condenser, and possibly the boiler, visible above the waves, i.e., *Alkimos* will be entering its terminal stage.

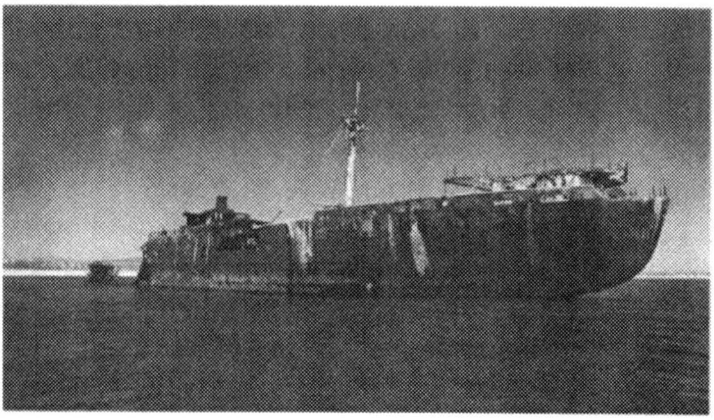

Figure 43a. View of SS *Alkimos* c. 1990, showing an intact stern section with extensive corrosion in the splash-zone amidships and a separated bow triangle (Pat Baker).

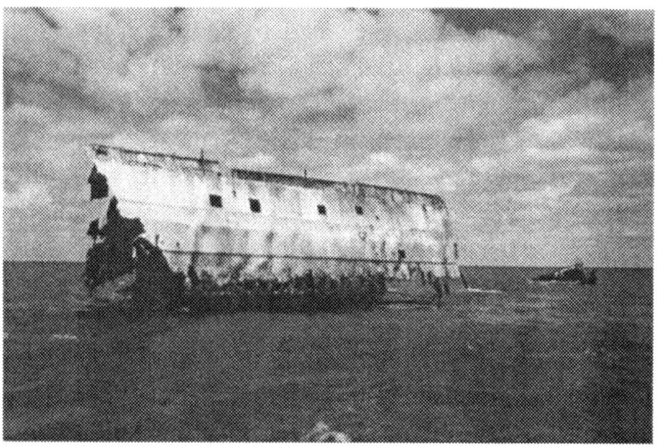

Figure 43 b. View of the SS *Alkimos* engine room in September 1999. The stern has collapsed into the sea (M. McCarthy).

The rapidity with which this particular wreck disintegrated, especially in the last five years, is remarkable. When compared with other Liberty Ships (Stewart, 1992), such as the SS *Elleni "K"* (1943–1966) in South

Australia (Arnott, 1999) and the *Cities Service Boston* (?–1943) in New South Wales (Nutley, 1992), helps provide a focus on the effects of mass-produced welded steel construction on site formation processes.

In respect of other variables (e.g., vessel size) in these processes, the wreck of the 500-foot-long (218 m), 10,925 ton, steel-hulled SS *Pericles* (1908–1910) provides an interesting contrast to its very well-known contemporary, the 883-foot-long (269 m), 52,310 gross tons RMS *Titanic*. Apart from the effects of its impact with the seabed and subsequent corrosion, *Titanic* appears to have sustained most damage to its structure on the surface and on the long c. 3,800 meter fall through the water column (Ballard, 1987).

In contrast, *Pericles* came to rest in 35meters of water on a hard reef floor, upright and virtually undamaged, bar a gash to its bilge—its bridge only meters below the surface. This once-massive ship has since totally collapsed due to the effect of huge seas and swell at the joining of the Indian and Southern oceans. The highest point today is one of its two triple expansion engines that now stands 10 meters proud of the seabed on a section of double hull. Plans of the vessel show that the top of this engine was once level with the ship's Plimsoll Line. Even at 35 meters down, three of the five boilers have rolled off their bearers due to swell. As the hull disintegrated, initially through natural forces and then in the mid-1960s through the use of explosives, the stern collapsed, leaving the two-cylinder compound steering engine and the rudder quadrant dominant aft. At the other extremity, the bow eventually collapsed and the walls of the chain locker also disintegrated, leaving a massive triangular mound of concreted chain lying on its side.

There are many examples of wrecks reaching the advanced stage represented by the *Pericles* and *Alkimos* cases. Articles appearing in the IJNA contain examples of this phenomenon in the waters off India (Gaur *et al.*, 1998) and elsewhere throughout the world (e.g. Barto Arnold, *et al.*, 1999; Watts, 1988). Examples also appear in the growing body of electronic sources pertaining to iron and steamship studies, e.g., the American blockade runner, PS *Denbigh* (1860–1865), a wreck currently being examined by a team led by J. Barto Arnold of the Texas A&M University (http://nautarch.tamu.edu/PROJECTS/denbigh/denbigh.html).

A progression from the situation represented by almost intact deep-water sites like *Titanic* toward the situation at the relatively shallow-water *Pericles* is seen in the wreck of the 665 gross ton SS *Blackbird* (1863–1878), shown in Figure 44.

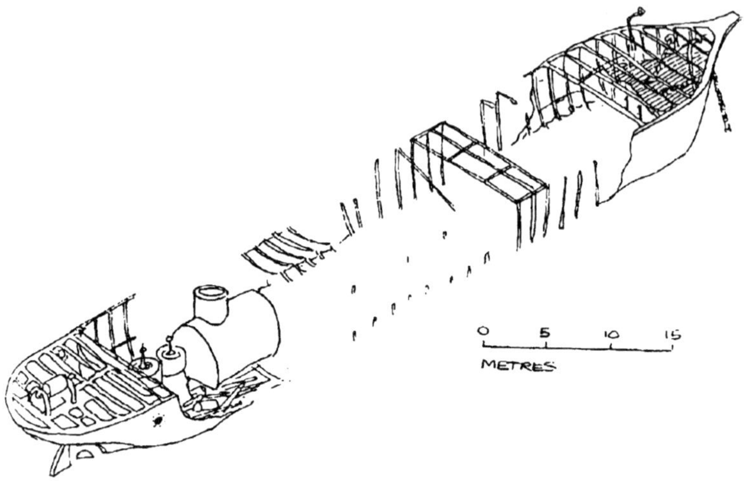

Figure 44. The SS *Blackbird* (Geoff Hewitt, 1988).

In ensuing decades the *Blackbird* site will be characterized by isolated vertical bow and stern triangles and a collapsed hull section surmounted by the engine and boiler. Then the triangles will collapse and will be found lying on their sides. Eventually the shell of the boiler will disintegrate and the engine's bearers will collapse, allowing it to fall from the vertical. Then only the engine and the two "triangles" will remain proud of the seabed, albeit on their side. The process of disintegration will continue, but for the lifetime of the present readers, this situation could be considered the "terminal stage" in iron steamship disintegration processes.

5.5 The formation of the *Xantho* site

It can be seen from the previous discussion that with two exceptions *Xantho* fits the patterns expected of an iron wreck lying submerged on a yielding seabed in wave-affected water for over a century. It appears to be buried to its waterline, the bow and stern triangles are essentially intact and upright, and as expected most of the hull amidships has disintegrated.

Contrary to expected patterns, a section of hull forward of the boiler is still standing proud of the seabed. This, we now know, is due to the fact that this section was exceptionally heavily built as a paddle sponson bearer. Further, although the wreck lies in shallow water opposite a break in the barrier reef protecting Port Gregory, the boiler has remained *in situ* and has not rolled out of the wreck as expected. This contrasts with the situation at *Xantho*'s contemporary the SS *Georgette* and this is possibly due to the protective effect of the sandbar along the port side of *Xantho*. It also suggests that the seas and swell at the *Xantho* site have not been as destructive as first thought. It is now evident, for example, that though the southwest seas impinging on the wreck refract around the outlying reef and strike the wreck from starboard to port have had some effect, their force is broken by the outlying reef.

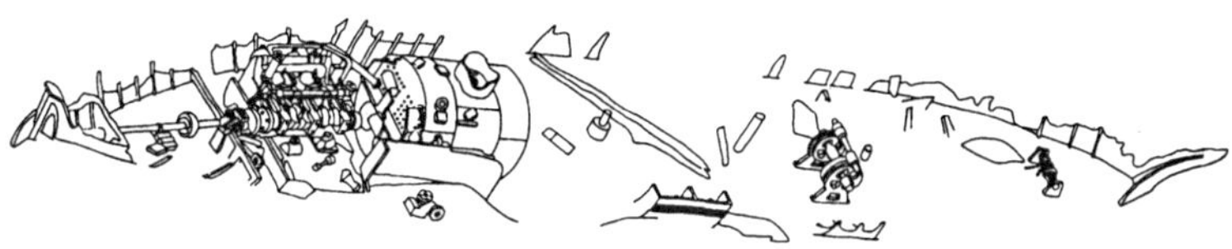

Figure 45. An isometric projection of the SS *Xantho* by John Riley. The original paddle sponsons and a coal bunker are evident on the starboard side. What appears to be a dog clutch is visible on the shaft.

By coming to lie on a mobile sand bottom in a surf zone and in the path of an overriding strong current running diagonally across the site from port bow to starboard quarter astern, the *Xantho* has opened up over the years to starboard. Being exposed to the full force of the current and the offshore movement of the sand bar, the port side of the ship has collapsed inward. Being in aerated water directly in the path of the current, the port sections have experienced a corrosion rate in excess of that found on the starboard side. The moving sand has caused the wreck to break up into four distinct parts, i.e., forward of the boiler, under the boiler, under the engine and aft of the engine. This disarticulation appears to have been solely due to natural forces.

On the other hand, the position of the wreck at the entrance to Port Gregory caused it to be a navigation hazard. The safety valve and steam dome, once atop of the boiler, have either been knocked off by a passing vessel or, being hazards to navigation, were deliberately torn from their housings and allowed to fall to the seabed.

An attempt to control for postdepositional effects and to come to an informed opinion as to why the wreck appears as it did in 1983 has been made. It is now relevant to ask whether behavior consistent with the circumstances of a vessel experiencing sudden hull failure and then going down in an unsuccessful attempt to return to port is indicated by the remains at the site. The answer appears to be yes.

The rudder is hard to starboard as would be expected of a vessel that abruptly struck a sandbar on its port quarter. The fire doors of the boiler are firmly shut, indicating that they were deliberately closed as the vessel slowly succumbed to inrushing water.

From the material evidence, it appears that abandonment was slow and that the grounding was accidental; i.e., the vessel was not deliberately run ashore for insurance purposes or for some other gain. There is independent confirmation of this interpretation from the testimony at the inquiries held after the loss of the vessel (Table 3).

Further, what occurred after the vessel sank was not the expected form of primary or secondary salvage that is the norm on most accessible steamship wrecks. The engine, pumps, and boiler remain, where normally these would have been removed by the purchasers or their agents soon after the vessel sank. Most of the copper and brasswork also remains where normally this is removed by modern salvors and sports divers.

Having accounted for both cultural transformations wrought by primary and secondary salvage and by natural transformations in the form of corrosion, current, seas, and swell, it is now possible to recognize anomalies in the material record and to account for them through archival and other analyses.

The *Xantho* has presented identifiable engineering and technological inconsistencies. It was powered by a horizontal trunk engine, a type expected only in a naval context, and steam was provided by what was apparently a slightly elliptical boiler of an early type. These were not the vertical compound engine and Scotch boiler that could be expected. A "condenser" was visible, although it did not appear connected to pumps, as expected. These anomalies need be accounted for.

Before doing so the author's projected stages of the disintegration of *Xantho* over the years are presented by Ian Warne in an inferred continuum that takes the reader from 1872 through 1979, when the wreck was first found.

All but the last image are conjectural and the compilation is based on observations discussed in this chapter and these in turn are the result of numerous iron and steamship site inspections over many years (e.g., McCarthy, 1996; Riley 1988a).

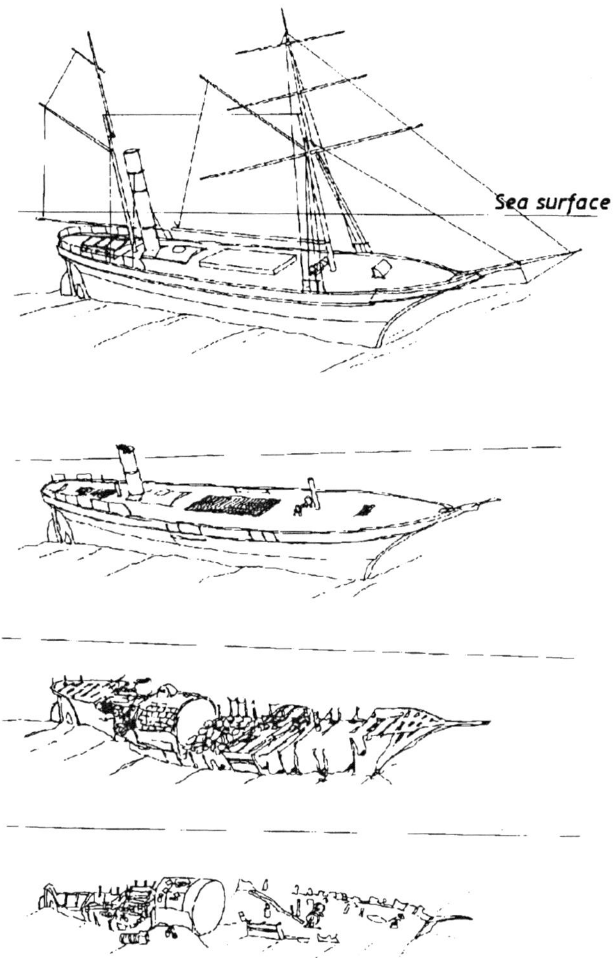

Figure 46. An impression of the stages in the disintegration of *Xantho* (Ian Warne).

Chapter 6

The investigation continues in the archives

The complementary and conflicing nature of the archival evidence

6. THE ARCHIVAL AND MATERIAL RECORDS AT ODDS

Some of the conclusions about the material remains on the wreck were not supported by subsequent archival research. By virtue of its apparently elliptical shape, for example, indications were that the boiler was an early low-pressure type and not the cylindrical Scotch boiler that became common in the later 1860s.

On the other hand, the local press described the boiler as being new when it was fitted to *Xantho* in 1871. It was also described as being multitubular, working to a pressure of 50 p.s.i., and manufactured by Davidson & Co., Boiler Makers, of Glasgow (*Herald*, January 25, 1873). This firm operated the Union Boilerworks out of Union Street, Glasgow, from 1867–1871 (*Glasgow Post Office Directory*, 1872). The boiler had to have been constructed between those years, casting doubt on the earlier conclusions.

The pumps were also described as new (*Herald,* January 25, 1873). These appeared to be part of an amorphous mass on the foreside of the engine itself and were, if those impressions were correct, an integral part of it. Being located in the stern of the vessel, they could not have been easily deployed to clear water from the bows or cargo spaces without a considerable amount of piping, however. These impressions fitted the evidence given at the court of inquiry into the loss of *Xantho*. On the other hand, the specifications for the building of the ship and the court of inquiry evidence contained references to watertight bulkheads. Yet the inquiry showed that water flowed from the bows aft and extinguished the boiler fires when the vessel struck the sandbar at Port Gregory.

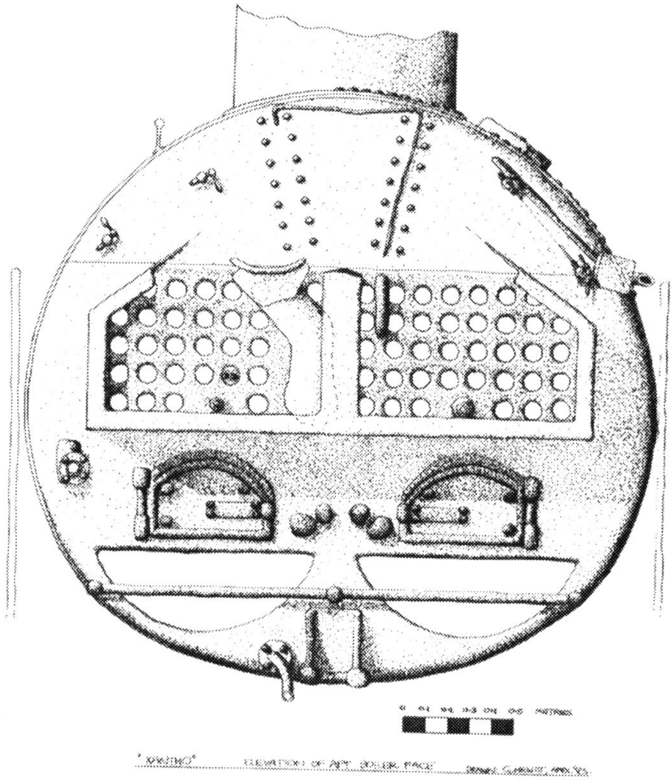

Figure 47. The aft face of the *Xantho* boiler (Geoff Hewitt).

6.1 The material record incorrectly read

Condensers were to be expected on a vessel built or refitted for use at sea in the early 1870s, as indicated earlier. All available illustrations of trunk engines show that they were fitted with condensers and that these were located adjacent the engine on the opposite side of the engine room. They were normally driven by the engine itself, but being located on the opposite side of the vessel, long pump-rods were required (see Figure 6). The location of what was originally thought to be a condenser directly opposite the engine, where it had apparently been attached to the starboard hull, corresponded to these expectations.

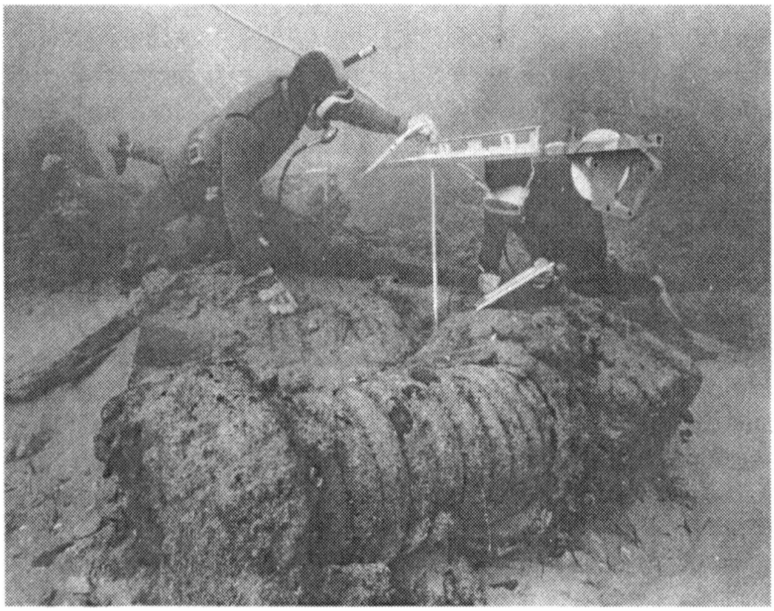

Figure 48. The object believed to be a condenser on the *Xantho* wreck (Pat Baker).

On the other hand, the expected pump rods, or their remains, were not visible on the *Xantho* site, posing another unresolved problem. Advice on trunk engine configuration was received from various sources, notably Noel Miller, our steam engine adviser, Joe Roone of the Science Museum in Kensington, and Richard Tomlin, a member of the team refitting the HMS *Warrior*. This famous museum ship was built in 1860 and was powered by a 1250 nhp Penn trunk engine. As the *Warrior* had been fitted with a trunk engine, which no longer existed (Warsop and Tomlin, 1990), a professional research assistant (Antonia MacArthur) had been employed to exhaustively comb British holdings at the National Maritime Museum in Greenwich for evidence of the engine type. This led to their accumulating a mass of information about the trunk engine, including descriptions, drawings, and photographs of models—notably those held at the Science Museum.

The consensus of opinion was that a large amount of cylinder lubricant was required for a trunk engine, and this would have precluded the use of a surface condenser. This was because tallow or animal fat lubricants would have insulated it and rendered it ineffective. It was generally agreed that *Xantho* should have had a jet condenser. Further examination of the drawings and photographs from the wreck suggested that the unit was unlike condensers of either the jet or surface type, however.

Given the absence of rods to drive the condenser pumps, it was reasoned that the *Xantho* may have had a noncondensing engine and that it exhausted to atmosphere, like a common steam locomotive. This was most unexpected, for the absence of a condenser meant that saltwater had to be used to top up the boilers. The 50 p.s.i. boiler pressures on *Xantho* (quoted in the newspaper report above) would have caused heavy precipitation, insulating the heating surfaces and requiring more coal to attain boiling point. As a result, coal consumption could be inordinately high. After some deliberation it was eventually decided that the "condenser" on *Xantho* was most likely a feed water heater, designed to warm cold seawater before it was fed into the hot boiler.

This conclusion was not easily reached. As the contemporary engineers Main and Brown noted in the quote reproduced earlier, a noncondensing engine will only be used for very short voyages where economy is not an issue, where space and weight (and hence draught) are at a premium, and where fuel is readily obtained (1855). These were not the parameters under which *Xantho* operated.

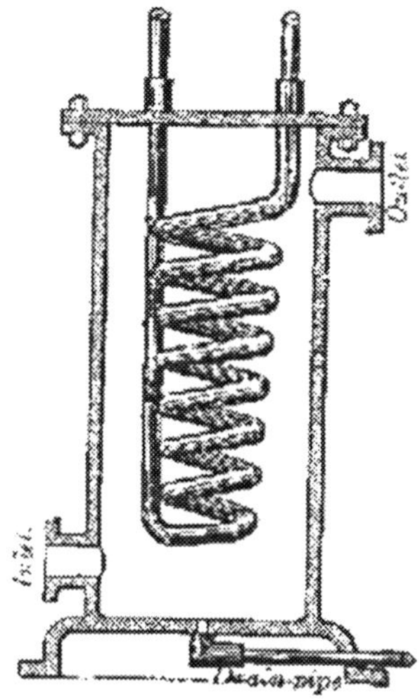

Figure 49. A feed water heater (Hutton, 1903:523).

6.2 Distortions in the written record

Attempts to identify the type of trunk engine located at the wreck also posed problems. It was recorded variously as 30 hp, 33 hp, 40 hp, 50 and 80 hp (*Liverpool Underwriters' Register*, 1871; *Inquirer*, February 8, 1873; *Herald*, January 25, 1873), for example. All agreed that the cruising speed of the vessel under steam was 7 knots, though there was disagreement as to the suitability of the engine for use in the ship (e.g., *Inquirer*, May 29, 1872; *Herald*, January, 25, 1873). The engine was also described as being only 18 months old and as "good as new", made by "Payne [sic], the Government machinist, of London" . . . a "masterpiece of workmanship . . . originally intended for a "Government gunboat" (*Herald*, 25 January, 1873). More concrete evidence was required for these claims could have represented an example of organizational behavior (Potter, 1992) in creating what eventually becomes archival material. In the first instance, the press would be expected to exaggerate, being motivated by their desire to see the ship succeed as the colony's first coastal steamer. In the second, the press were evidently keen to see the wreck of the *Xantho* sold off for the best possible price in order to pay the arrears of wages owing to the crew when Broadhurst abandoned them.

Further, in the 1871 *Xantho* register (Figure 21), the engine was described as a horizontal engine, built by Penn and Son of London in 1861, with 20-inch diameter cylinders and a 13-inch stroke. While originally recording it as a 30 hp engine, the description was amended two weeks later to read 60 hp. This alteration helped resolve a number of problems.

The reasons for the confusion stem from the fact that the engine type was not specified and (as in most cases) only the barest details are given in the registers. Further, the word *engine* is often used to refer to an individual cylinder. Thus *one engine* can mean one cylinder and the term *engines* can indicate two or more cylinders in any one piece of machinery. In the *Liverpool Underwriters' Register* (1872:381), for example, *Xantho* appears as having "one horizontal 30 hp engine" and in some of the colonial newspaper accounts reference is made to "these" engines. Each cylinder of the *Xantho* engine apparently developed 30 hp and the two cylinders combined produced 60 hp in total. The 1871 Glasgow register that originally carried the description "one 30 hp engine" was subsequently altered, possibly in recognition of the fact that the engine had two cylinders.

6.3 The naval origins of the *Xantho* engine

By combining the written sources with the evidence from the wreck the engine was finally identified as a noncondensing, two cylinder 60 hp trunk engine built in 1861, by Penn and Son of London, most likely for the Royal Navy. The ability to resolve the difficulty experienced here attests to the value of the artefact as a primary historical source in its own right and it clearly illustrates both the complementary and conflicting nature of the documentary record. The perils of accepting the written word *verbatim* are clearly indicated here. So too is the need to take considerable care where the archaeological record is the sole source of information, especially where the required range of specialists (e.g., in this case–steam engineers) are not present. There is also room for debate on the extent to which the documentary and material evidence is expected to balance or complement each other, or even which of the two takes primacy and in what circumstance that occurs (e.g., Deetz, 1988).

These conclusions and Geoff Kimpton's drawings of the *Xantho* engine in its concreted state were also analyzed by the *Warrior* team. Though they could not find any drawings of the type, they advised that the unit appeared similar to those described as the 40 or 60 nominal horsepower (nhp) trunk engines built by John Penn for the Crimean War gunboats of 1854–5. The latter had a cylinder diameter of 21 inches, a stroke of 12 inches, and a computed trunk diameter of 11 inches, producing 60 hp. They developed 270 indicated horsepower (ihp), which drove the gunboats at approximately 7 knots (Preston and Major, 1965; Osbon, 1965). See Appendix 1 for an examination of horsepower and its derivatives.

The engines were noncondensing because the gunboats were designed to have a very shallow draught to enable them to safely negotiate shallow waters, as a matter of strategic priority. This required that weight be kept to a minimum—and condensers were heavy. (See Main and Brown's comments above.) Being compact vessels, with gunnery the other main priority, space was also at a premium. This was another consideration mitigating against the fitting of a condenser.

The primary factor that enabled the designers to dispense with the condenser in their efforts to save weight and space and thereby attain the required operating parameters was the nature of the Black Sea, their intended field of service, however. It had a salinity of 16–18 parts per thousand, being only half that of the major oceans of the world (USSR Ministry of Defence, 1974; Florian, 1987). The salinity of the ocean around the Western Australian coast was double that of the engine's original intended operational conditions.

A contemporary source noted that these 60 hp, noncondensing, double-acting, double-trunk engines exhausted to atmosphere and drove a two-bladed Smith's screw measuring six feet (1.8 m) in diameter (*The Engineer*, February 11, 1898). It appears that over 150 sets of 60 nhp engines were built for gunboats in the course of two years; half by the firm of John Penn of Greenwich and half by the firm of Maudslay Son and Field of Lambeth. Penn used his well-known trunk design; Maudslay, the return-connecting-rod type. Both allowed for the compactness required of an engine designed to be kept below the waterline away from shot and shell. Of additional interest was that some form of subcontracting was needed for such a large order in such a short space of time. As a result, it appears that engineering firms elsewhere produced items such as the cranks and connecting rods and that, on delivery, the two firms completed only the final assembly and installation phases (Preston and Major, 1965). As a result, the Crimean War gunboat engine is recorded as "probably the first recorded instance of mass production being applied to marine engineering" (Preston and Major, 1965; Osbon, 1965). Thus, on the basis of historical evidence it appears that Penn and Son need be credited jointly with Maudslay Son and Field for producing the first mass-produced marine engines in the form of the Crimean War gunboat types.

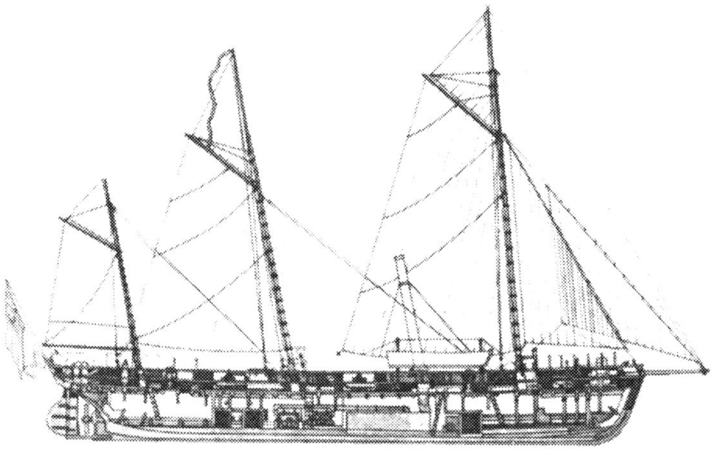

Figure 50. A Royal Navy gunboat of the Crimean War type (Archibald, (1968:89). Illustration Clifford Meadway.

As with the large trunk engines, there was apparently a very high standard of workmanship in the construction of the relatively small gunboat engines. Not only were high standards a hallmark of Penn's workshops, but the gunboat engines represented a watershed in marine engineering–operating at high speed (190 revolutions per minute). This was an enormous leap considering that other marine engines of the time, such as that fitted to HMS *Warrior,* were rotating at only 60 revolutions per minute at most. They were also high-pressure engines, using steam at 90 pounds per square inch, or six atmospheres (Bar). This represented a quantum leap forward both in the pressure applied and in the speed of the engine (Preston and Major, 1965; *The Engineer*, February 11, 1898).

The success of this type of engine saw the end of the "foolish prejudice against the use of high pressure engines or boilers at sea" (*The Engineer*, February 11, 1898). They also heralded the removal of restrictions placed on steam that, up to 1853, kept it to 30–45 pounds pressure or 2–3 atmospheres in British service. It was considered an "enormous leap" when compared with practices of only a few years earlier. These engines, described as having cylinders "no larger than that of the land locomotive of the time," were apparently the first high-pressure, first high-revolution, and first mass-produced marine engines made to be used at sea (*The Engineer*, February 11, 1898).

The problems created by the high pressures and high speed (e.g., excessive wear and overheating of bearings) forced the Royal Navy to keep a floating workshop on the "China Station" to service the later engines (Preston and Major, 1965:108). Despite their limitations, some units remained in service for many years and a number lasted into the late nineteenth-century and beyond (Osbon, 1965). Despite this, the problem of heat loss from the exposed trunks, wear from ash and other abrasive substances, power losses caused by friction at the trunk packing glands, and the associated large appetite for coal (even when fitted with a condenser) were substantial. These served to see the trunk engine, no matter how well engineered, uneconomic in comparison to the compound engine developed in the 1860s (Guthrie, 1971). Unfortunately, plans and detailed descriptions of the gunboat type have not yet been found.

Thus the 23-year-old, former paddle steamer *Xantho* had been fitted with a 10-year-old gunboat engine during the refit that took place at the hands of the scrap metal merchant Robert Stewart. The first questions requiring answering were how or why did this occur and what was its significance in general terms?

The answer was found firstly in the shortage of seasoned oak that occurred in the mid-1850s. This resulted in permission being given to the builders of the 1854–5 gunboats to use a variety of woods, some of which were known to be green–i.e., unseasoned and liable to warp (Osbon, 1965, Preston and Major, 1965; Archibald, 1968).

The gunboats ordered for the Crimean conflict and later posted to foreign stations (for example, China, the Mediterranean, North America, the West Indies) achieved notoriety when some of them literally fell apart within years of their launching. Even those built after the Crimean War were in some cases hastily constructed. A few years later, for example, three-quarters of a large group of vessels found rotting away in shipyards were the 60 hp gunboats and there was a great scandal as a result (*The Engineer*, January 3, 1862). When these gunboats were being disposed of, their machinery had seen little service, and 56 engines were fitted into the larger Plover class of twin-engine gunboats (Preston and Major, 1965). Recycling of gunboat engines then became the norm.

Assuming the 1861 date for building (or final assembly) of the *Xantho* engine is accurate, and given the slightly different 20-inch cylinders and 13-inch stroke recorded in the *Xantho*'s 1871 register (Figure 21), it appears that it may have been part of a subsequent batch of gunboat engines.

Of the Britomart class of post-Crimean War gunboats, a total of 20 were ordered; 10 in 1859 from private shipyards and a further 10 in 1861 from Portsmouth. These included the vessels *Bramble, Crown, Danube*, and *Protector,* which were laid down (or ordered) in 1861 and cancelled on the stocks (building yard) on December 12, 1863 (Colledge, 1969:14, 87, 152, 441). They were 120 feet (36 m) in length, 330 tons, and had wooden hulls. Though not conclusive, the chances of the *Xantho* engine coming from one of these four vessels is reasonably high, given the date of manufacture in *Xantho*'s register is 1861 (Figure 21). Equally, where mass-production techniques were used, *Xantho* could have had an engine that was built earlier and stored in parts until required in 1861.

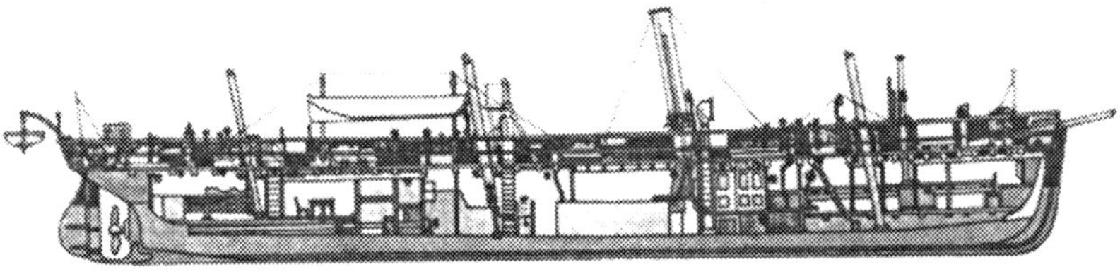

Figure 51. A *Britomart* class gunboat (Archibald, 1968:89). Illustration Clifford Meadway.

6.4 Broadhurst revisited

Having accounted for the wreck, the materials found on it, and (to an extent) their antecedents, the behavior of those involved came into focus. In examining this instance of nineteenth-century recycling, the question arose whether the scrap-metal merchant, Robert Stewart was dishonest–a marine equivalent of today's crooked used car dealer? These, indeed were my original thoughts (McCarthy, 1985, 1986a).

In highlighting the difficulty of attaining agreement on motivation and explanation for behavior it should be noted that Stewart may not have been dishonest. He may have assumed that a prospective buyer would operate the ship around Britain, or on the continent, in sheltered waters, close to engineering, repair, and coaling facilities. Given that the ship was old and originally designed for use in inland waters, it was a reasonable expectation that it would be operated where the stress on the hull would be kept to a minimum. Stewart may also have assumed that it would be operated in waters that were fresh, or nearly so, and over short distances to compensate for the lack of the relatively expensive and bulky condensers. He probably never dreamed that it would travel across the world and then lie on tidal flats in the remote north of Australia. In support of this assessment, and as indicated earlier, the 1871 register recorded the *Xantho*'s carrying capacity in both tons and cubic meters. This was apparently with a view to its sale to either British or continental interests or at least to countries such as France, where the metric system was in operation. Though freshly painted, with increased cargo space resulting from its conversion to screw propulsion and with efficient cargo-handling facilities, the vessel's age, comparatively small scantlings, and its mode of engineering could not have escaped the attention of any knowledgeable buyer.

The intent to defraud was not necessarily in Stewart's mind. On the other hand, perhaps he did not care and his interests were only on profit. In recognition of the latter possibility it was proposed in 1988 that "the refit, re-engining and sale of the SS *Xantho* in 1871 was not in the best interests of the unfortunate purchaser" (McCarthy, 1988a:347).

It was also believed that Broadhurst's decision to purchase *Xantho* was ill conceived at best and his behavior was evidence of naiveté and poor judgment. That the ship appeared on the very distant Western Australian coast naturally posed the question: what sort of person would purchase it for use on the sparsely populated, poorly serviced Western Australian coast, far from the nearest marine engine repair facilities at Adelaide and Surabaya? Was he an impulsive individual with vision and access to capital yet lacking the necessary experience of practical, frontier capitalists? Worse still, was he a rich gentleman eccentric (a fool even), given to grand and ill-considered schemes, as first thought?

Thus, a fundamental change of emphasis occurred after the inspection of the site and attention came to focus on how Broadhurst came to make the apparently strange decision to purchase this odd hybrid refugee from the scrap heap (McCarthy, 1985). By shifting emphasis to include the social context, the project ceased having the purely descriptive focus of traditional maritime archaeology and came to have an additional analytical/explanatory element, evolving into the *Xantho*/Broadhurst project as described in these pages.

In that context, it was considered necessary to examine Broadhurst's career both pre- and post-*Xantho* in order to ascertain whether a consistent pattern emerged and whether further clues were evident there. Though the research conducted was exhaustive and detailed (McCarthy, 1990), at this juncture it suffices only to provide a very brief overview and then the results.

6.4.1 Pearling at Shark Bay

Broadhurst initially intended to work the Shark Bay, Nickol Bay, Flying Foam Passage, Port Hedland, and Banningarra pearl shell fisheries shown in Figure 13 simultaneously. All of these places, especially Shark Bay, had confined, shallow, and difficult waters that were a trap for sailing vessels but ideal for a small shallow-draught steamship. When the *Xantho* (which was to be the link between these centers) sank, Broadhurst was placed in a difficult financial position without the means to link his widespread endeavors. By having allowed his insurance on *Xantho* to lapse, he did not have the funds to pay his men or to purchase another steamship, let alone a suitable sailing vessel with which to re-establish the links. On the other hand, he had a large workforce of Malays at his disposal and was able to muster some small boats that would prove suitable for shallow water work. Given the problems resulting from the loss of the steamer, which was his linking mechanism, the answer lay in concentrating at one location. Shark Bay was logical, for there the pearl oyster (which was noted more for the pearls that it contained rather than the shell) was easily harvested and Broadhurst's divers (who continually failed) could obtain it without even getting their feet wet (see McCarthy, 1990).

6.4.2 American influences

Broadhurst was the best-organized operator and had at least 7 small cutters between 1–2 tons each: *Shenandoah, Alabama, Florida, Talahassy* [sic] *Stonewall, General Lee,* and *Jefferson Davis.* One boat, named after his illustrious British brother-in-law, *Sir Joseph*, was of an unknown type and size. He also had two other pearling vessels, the dinghies *Xantho* and *Pearler.* The reasons behind his propensity for names from the American South is not known. There is certainly more than a coincidence in it, for it was an era when American enterprise and entrepreneurial flair were generally seen as worthy of emulation. Many years later, for example, Broadhurst was described as having "all the decisive action of a typical American" (Kimberly, 1897:8).

6.4.3 Broadhurst: A great success

Initially Broadhurst was a great success and he was eulogized in language reminiscent of the effusive tones used to describe the heroes of a gold rush:

> And now we are startled by the announcement of one gentleman–Mr. Broadhurst–receiving as the proceeds of one month's fishing no less than 100 ounces of pearls worth at least £2500. Than Mr. Broadhurst no man better deserves his present success. He has been an energetic speculator, undaunted by adverse results (*Inquirer*, October 8, 1873).

This unaccustomed success was due to both the number of men at Broadhurst's disposal and the fact that he had a very capable manager who spared "neither labour nor trouble" in improving the condition of the Malays in regard to their housing, health and work practices (*Inquirer*, October 8, 1873, October 29, 1873). Another factor was the technology employed. It was arguably simple, being small wooden sailing vessels dragging a triangular dredge in which the shells were harvested.

Evidently Broadhurst's primary mode of operation was to set up a base himself, work there for a while alongside his men, sharing their risks (he nearly drowned in one instance), and then appoint a manager. He would then set off in search of other enterprises and activities. Such was the case in Shark Bay, where he not only diversified his maritime pursuits but also established a wooden store, thereby profiting from the sale of food, goods and liquor. This was managed by Eliza's nephew. Broadhurst also submitted a tender to sink wells on the stock route south from Shark Bay and established "fishing stations" throughout the bay. He then sent trial shipments of the supposedly worthless pearl shell to Europe and while awaiting news of the venture, the following optimistic comments appeared in the press:

> The speculation is a bold one . . . and on Mr. Broadhurst's account alone, who for enterprise and determination is a man out of 10,000, we sincerely hope the venture will turn out as profitable as Mr. Broadhurst could desire and make up in some measure for the heavy losses he has sustained in other undertakings (*Inquirer*, September 2, 1874).

In this comment we find that Broadhurst "a man out of 10,000" was on his way to being socially resurrected as a colonial benefactor after the Denison Plains fiasco, his seemingly endless failures, and his much-publicized failure to pay the crew of the *Xantho* when it sank.

6.4.4 Broadhurst: The consistent failure

After these initial successes, again things started to go drastically wrong; this time following the replacement of Smith with Broadhurst's nephew, Daniel. It could not have come at a worse moment—for Charles Broadhurst had just been nominated to represent the northwest pastoralists and pearlers in Parliament, having been considered as "the only person in the position who has the abilities and leisure to represent them" (Drake-Brockman, 1969:234). He sat in the Chamber for the first time in late October 1874 but was forced to resign almost immediately when a scandal broke about his nephew's failure to pay and then repatriate the Malays. Daniel Broadhurst left them wandering around Shark Bay, destitute, and as a result Charles Broadhurst and his entire family received another severe social and financial setback. Summonsed six times in Perth during this period for non payment of debt by a cross section of the European merchants in the colony, his finances and reputation were obviously in a very bad state. He then left the colony with a second cargo of pearl shell, arriving back in April 1876, with the sad tale that only the middleman made anything out of the venture. Broadhurst's fortunes and morale were at such a very low ebb that he applied unsuccessfully for a position as a Sheriff in the public service. Forced to make the most of any opportunity, he then left for London with yet another cargo of Shark Bay shell.

6.4.5 Broadhurst's other enterprises

Eliza was not to see her husband for over a year. When he did return in October 1877 his mind had turned to canning fish. Somehow, Broadhurst interested a financial backer, established buildings, erected machinery, and set up the Mandurah Fish Canning and Preserving Company, just south of Fremantle. When at full capacity, the works employed about 50 people from a total population in the town of 200. It proved moderately successful, but ever-restless and searching for opportunities, he sold the business in 1882 and proceeded to the Abrolhos Islands near Geraldton (Figure 13). There he discovered previously unknown and very extensive beds of guano, a rich organic substance consisting of the remains of birds, their droppings, and other material. It was much sought-after throughout the world as a fertilizer of great significance (Stanbury, 1993). In this display of creativity, Broadhurst clearly deserved being later described as "a capitalist and trader who would go out of the ordinary grooves in search of wealth" (Kimberly, 1897:90).

The government decided to grant him a lease to the guano deposit and in December 1883, at 57 years of age, at a time of life when most people would have ceased their struggle for security and wealth, Broadhurst settled at the Abrolhos Islands. He had difficulty in finding surety for the guano lease, however, and was forced to take on both a "silent" financial backer and then an active partner, forming the firm Broadhurst MacNeil and Company. At the time there was little profit margin in the business, and as a further indication that money was tight (or as further proof of his attitude toward creditors of all position, race, and creed), Broadhurst was summonsed for nonpayment of debt by a cross-section of the colonial business community. His former backers in the fish-canning industry also summonsed him.

At 44 years of age, Eliza was struggling to make ends meet in the face of her husband's continued financial woes. An advertisement announced her intention to open a girls school and a preparatory school for boys under 12, with a few boarders at a "reasonable rate" (*West Australian*, March 4, 1884). She was noted as being ill and looking "quite fagged and worn out" from school and music lessons (Hillman diaries, October 7, 1883). Life was certainly tough for them both and they still did not even own a home, for in January 1884 Eliza sought to rent a cottage for £50 per annum plus rates and taxes (Hillman diaries, January 21, 1884).

The scene changed dramatically in 1886 when their son Florance Constantine Broadhurst became a business partner in the company. Recognizing his own deficiencies as a business manager and seeing that his son, who had received a mercantile education and had been a success in the banking industry, was to be the key to the future success of the business, Charles Broadhurst retired. Under Article 9 of the agreement under which Florance was brought into the company, it was stated that

> The management of the firm shall be exclusively in the hands of the said William Brown MacNeil and Florance Constantine Broadhurst as joint managers . . . and the said *Charles Edward Broadhurst shall in no way interfere in the management of the said partnership business* (Broadhurst Family Papers).

The agreement and his son's success are clear indications that Broadhurst senior, though capable of seeing opportunities and grasping them, was a poor administrator. A further indication of this is the statement that proper books were to be kept and that a general accounting was to be made at the end of each calendar year. Thus it was not Charles Broadhurst who turned guano mining to his advantage, but his son Florance, a man later noted for his "clearsighted methods and organising power" (Kimberly, 1897:98; *West Australian*, October 24, 1887).

Through the sale of guano the family became extremely wealthy, and in May 1890, at the age of 64, Charles Broadhurst retired from the firm. A family trust to the amount of £10,000 was then formed in order to secure the future of Eliza and their children and so they returned "home" to Bournemouth in England where they established *Karrakatta,* their family seat. In 1899, Eliza died at age 59, apparently worn out from her exertions, leaving an as-yet unwritten yet truly remarkable, story to be told (McCarthy, *et al.*, in prep.) When Charles died in 1905, at age 79, news of his death was quickly relayed to the former colonies. He received a suitable eulogy in the Victorian and Western Australian press, which in summing up his career, read as follows:

> Mr. Broadhurst was one of the most indefatigable and persevering exploiters of the infant industries of Western Australia in his day (*West Australian,* May 1, 1905).

Even from the grave he was to cause confusion and concern. The executors of his will found themselves unable to fulfill the conditions in the manner specified due to there being insufficient funds to attend to his wishes–indicating once again and for the last time his lack of attention to detail in such matters.

Chapter 7

Excavations at the site

7. EXCAVATION PROCESSES

In the manner described in the previous chapter, the author had come to the realization that Broadhurst was a significant colonial entrepreneur and that the *Xantho* was of considerable historical and technological importance. Further, its engine was not just as an intact example of the rare trunk engine type, but was an example of one of the earliest high-pressure engines and the first mass-produced, high-revolution engines used at sea. It was physically degrading and also was in danger from recreational divers even though anodes had been applied and it was camouflaged with rocks and sediment to keep looters at bay.

7.1 The research strategy

The possibility of raising and then conserving the entire engine in the context of an excavation centering on Broadhurst was discussed with Neil North. He indicated that, if raised, it would have to be treated as one piece and that the main problem was size. North also indicated that if the engine were to be raised and placed in a treatment tank with appropriate infrastructure, the external surfaces of the engine could be readily deconcreted and treated by electrolysis, then a common technique being successfully applied to large iron objects (North and MacLeod, 1987). He felt that the internal cavities in the engine would be a much more difficult proposition and that the only feasible approach would be to clean them as well as possible and then electrolyze them using insulated "rod" anodes. He predicted this would be a slow process and that the main costs were the tank, the provision of space for the tank, treatment chemicals, and labor. Failing the provision of a tank, he believed that the engine could be left in its concreted state and placed in a safe underwater environment at a naval base near Fremantle. There a sacrificial anode system could be attached to protect the engine against corrosion.

When all of this was considered, and with the safety of the engine a prime concern, the author came to favor the option of raising it as an endangered artefact of considerable significance. The matter was taken to the Maritime Archaeology Advisory Committee (MAAC), a group of academics, practitioners and representatives of the diving communities which advises the museum's director on wreck-related matters. Their support was subsequently obtained (Maritime Archaeology Advisory Committee Minute Book, June 22, 1983).

The decision to assess whether the engine could be successfully raised, transported, and conserved for the purposes of exhibition and further study was relayed to Broadhurst's granddaughter, Marjorie Darling. She received the idea with enthusiasm, thereby giving a final seal of approval as the last to have a direct familial link with Broadhurst.

7.2 Preliminary fieldwork

The means of removing the engine from the wreck and of transporting it to Fremantle were assessed, with favorable results. The problems identified in this exercise were routine recovery procedures governed by object size and water depth. On the other hand, the weight of the engine and hence the number of lifting bags required was difficult to predict. Eventually a weight estimation formula was obtained from a former steam engineer—(estimated weight of an engine (in tons) is the product of its length, by breadth by height (in feet) divided by 27 and multiplied by 1.3. This produced a range for the *Xantho* engine of between 7–10 tonnes—allowing for the unknown weight of concretion (B. Docherty, personal communication, February 22, 1984). Then the author and Geoff Kimpton, a former oil-field diver, conducted a series of underwater trials in order to test the most archaeologically acceptable method of cutting the engine free from the wreck. A number of successful "cuts" were conducted on a modern wreck, the SS *Lygnern* (1920–1928). Though built of steel, in other respects (mainly depth and concretion layer) it emulated the situation at *Xantho.* Following these tests, it was eventually decided to use a thermal lance (an oxygen/steel-powered, high-temperature cutting tool) to free the engine.

7.3 Further site assessments

On arrival at the site in mid-January 1984, the team was startled to discover that the wreck was almost totally covered in sand. The engine was buried up to the top of its trunks.

The boiler, which nine months earlier stood almost clear of the seabed, had only 1.5 m of its upper surfaces protruding from the seabed. Apart from the top 20 centimeters of the sternpost, the remainder of the site was not visible. Despite this severe, yet informative, setback, the strong currents and poor general conditions present in the April 1983 study were not in evidence.

North and MacLeod again joined the team in order to examine the success of the anodes previously fixed to the wreck and to monitor the extent of biological regrowth in the nine months since the first predisturbance survey was conducted.

The newly deposited sand around the engine was cleared and the anodes were re-located after considerable effort. They were found to have had only a limited effect, due to their burial in sand and clay in the intervening months. After being excavated they were placed on a higher level on the understanding that they would begin protecting the engine once they were returned to an oxygenated environment. The sand cover had also effectively killed all marine growth on the engine and boiler, except the upper surface, leaving only newly colonized weed and dead molluscs on the concretions.

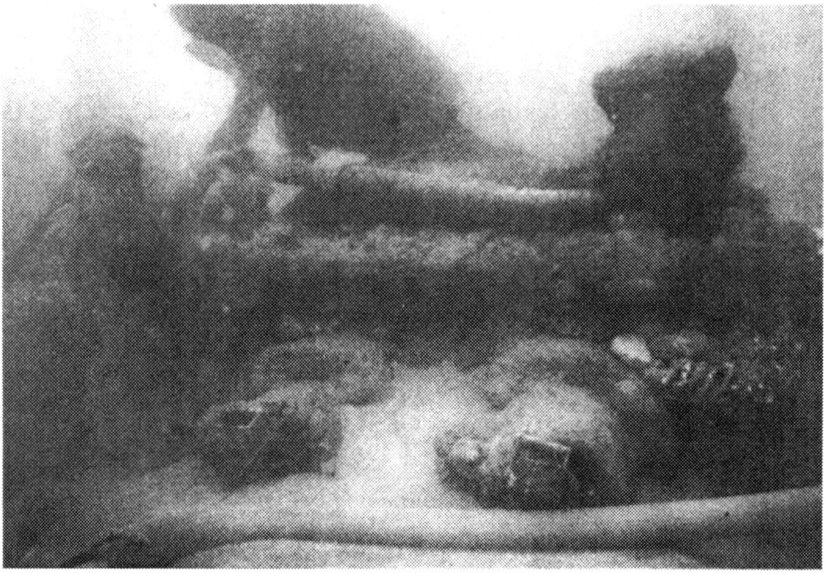

Figure 52. The extent of the sand cover on the engine in 1984. Compare the cover with the situation the year before in Figure 29 (M. McCarthy).

The burial appeared to be a recent phenomenon, for some of the dead bivalves attached to the boiler were still articulated. Here were clues to the bands of living and dead marine growth on the boiler in April 1983. North concluded that

> Sometime between May 1983 and January 1984 there had been a tremendous deposition of sand onto the *Xantho* site and this had [negated] nearly all our experiments. . . . My interpretation was that, for some reason, the site has been completely covered with sand so that none of the wreck protruded above the sea level. Approximately 3 weeks ago some of the sand was swept away to reveal the top 3 feet of the boiler and colonization started again. In the last 1 to 2 weeks a second sand shift has exposed the boiler for a further 2 feet. . . . We had a further disappointment with the anode system. The potential measurements indicated that it wasn't working and I felt this may

> feet below the present seabed we ran into a mixture of fine sand and clay. I estimated that there would have been 3 feet of this material overlying the anodes . . . and above that another 10 feet of sand . . . so it is not surprising they were not giving out much current (N. North to C.J. Beegle, personal communication, February 1984).

Having been filled with mobile sand since the 1983 program, the area around and underneath the engine was excavated in a gross manner (i.e., quickly and without recording). The space beneath was examined in order to gauge the size and strength of the engine bearers. There it was ascertained that the engine was bolted to a rudimentary box frame consisting of riveted plate iron, originally 1 inch or 25 mm in thickness and approximately 18 inches or 500 mm deep. The frame itself was riveted to lateral supports or beams of I-section iron, running across the engine room to each side of the ship where they were riveted to the hull and frames. Supporting the fore lateral bearer and fixing it to the keelson was a vertical iron plate measuring 1 foot (30 cm) wide and 1 inch (25 mm) thick. This was further braced to the keelson with a similar sized diagonal bearer. The lateral engine bearers were extremely fragile. To an extent this was expected, for though originally 1 inch or 25 mm thick, they had been exposed on both surfaces to the elements and corrosion was occurring on each side. At a predicted rate of 8–10mm per 100 years on both sides, it was expected that a minimum of 16–20mm of the original 25mm thick engine supports were severely corroded. On the other hand the vertical supports fore of the engine were underneath the structure and were thereby protected from the current and the seas. As a result they had substantially more original metal left. The extent of the corrosion also confirmed North's prediction that the engine would begin to disintegrate after a few decades, or at best it would collapse into the ship's bilge.

7.4 The engine cut free

After exploratory work in the bilge and around the engine, copper piping was found passing from the engine under an iron decking and into the bilges underneath. It also became apparent that due to the rudimentary construction of the engine bed and its advanced state of decay, very few points needing cutting in order to free the engine. A suitable location (for what was originally planned to be a test cut) was identified and marked, and large timber bearers were wedged under the engine should it collapse on the divers. There was so little original metal holding the engine and its box frame to the vessel, that within the space of a few hours the engine was free. After cutting a number of copper pipes with a hacksaw, the engine slowly settled downward onto the timber (Kimpton and McCarthy, 1988). Anodes were reapplied, the engine was again camouflaged and the team departed.

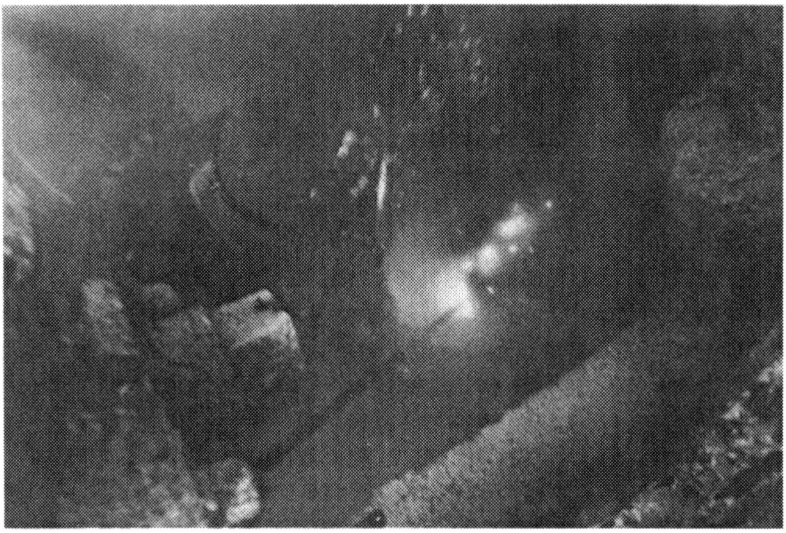

Figure 53. Geoff Kimpton cutting the *Xantho* engine with the thermal lance (M. McCarthy).

7.5 Budgets and rewards

Briefings on the 1983 and 1984 phases of the SS *Xantho* project were regularly made to the museum's Maritime Archaeology Advisory Committee mentioned earlier. In receiving various reports on the merits of the project, discussions were held on the question of a reward to the finders of the wreck. The justification for the reward took into account the vessel's pioneering role, its links to the Broadhursts, and its importance to the history of marine engineering (S. Sledge, to Director,

WA Museum, October 2, 1984). Letters of support from the HMS *Warrior* team and from the Water Transport Department of the Science Museum in London were also provided. These indicated that the *Xantho* engine was an "exciting and unique find," one which would be (in their words) "invaluable" in order to "fill in the gap" for a period were no records had been found (R. Tomlin to McCarthy, December 28, 1983; J. Roone to McCarthy, May 18, 1984).

Eventually it was decided to present to the Maritime Archaeology Association of Western Australia (the group that first reported the wreck to the Maritime Museum) a reward of AUS $3,000–the largest sum ever given under the terms of the State Maritime Archaeology Act of 1973. Until late 1994, when rewards were retroactively paid to the finders of the bullion carrying Dutch and English East-Indiamen, it was the third largest reward paid in Australia under state or federal shipwreck legislation. This was a fair indication of the changing perception that, though built of iron and relatively modern, the SS *Xantho* was one of the most significant wrecks found in the waters off Western Australia. It was also a major change in direction for research in maritime archaeology within Australia.

In addition to the reward, a budget of AUS $7,200 was allocated to the next phase of the project: the excavation of the stern and the recovery of the engine. Being a relatively small sum in comparison to that then allocated to the excavation of wooden wrecks, it was augmented with sponsorships, tax incentives, and other schemes.

Shortly before the expedition got under way, Neil North left the museum for private enterprise and Ian MacLeod became responsible for the subsequent on-site and in-laboratory conservation phases at *Xantho*.

7.6 The iron and steamship wreck seminar

Being a new element in Australian underwater archaeology, calls for expressions of interest in participating in both the excavation and a seminar on iron and steamship wrecks were made through the Australian Institute for Maritime Archaeology (AIMA). The seminar was to be the first on-site practical and theoretical seminar on maritime archaeology held in Australia.

The excavation and the seminar opened simultaneously in mid-April 1985. In providing a historical background for iron shipbuilding and steamship operation, in detailing the iron and steamship wreck resource throughout Australia, and in examining existing management programs, the seminar provided a useful framework for Australian iron and steamship studies at the time. Recording and conservation techniques specific to iron and steamship wreck sites were also explained in both a theoretical and practical sense, some overseas developments were considered, and numerous survey and excavation reports were received. Eventually 45 papers were published (McCarthy, 1988b).

As mentioned previously, one influential Australian maritime archaeologist Graeme Henderson, the head of the Western Australian Maritime Museum's colonial wreck program, challenged the seminar audience by paraphrasing David Lyon of the National Maritime Museum (Greenwich). With over 500,000 plans of ships in the collection, Lyon was generally concerned at the wisdom of spending great sums of money to make what could only be "generally inferior" plans of shipwrecks on the seabed, especially where concretion served to make recording even more difficult (Henderson, 1988:11). Henderson also echoed the commonly held belief that iron wrecks "tend to last better in many environments" and argued that they would not constitute what he termed, the "individual expression of creativity" which was to be found on the wooden wreck. Further, in acknowledging that there were "fundamental, as yet unexplained, questions about conservation and restoration techniques because of the past lack of scientific attention to such sites," he also suggested that "no further engines be raised at least until the *Xantho* engine has been successfully conserved" (1988:11). It was a pertinent and valid point, highlighting the experimental nature of the project and of iron and steamship wreck studies generally.

7.7 The excavation

The wreck site was found to be in similar condition to that found on the first inspection almost exactly 2 years earlier in May 1983. The sternpost, tip of the propeller, all of the engine, most of the boiler, deck winch, windlass, stempost, and sections of the port hull were exposed. A copper cable attached to an anode that had been connected to the engine was covered with a sulphide patina, indicating that it had again been buried under sediment. The anode itself was found to have been partly working, resulting in a reconstituted layer of calcium carbonate on the copper and brass fittings. The anode attached to the stern shaft also appeared to have worked, but again only partially, due to its having been buried. Corrosion potentials were recorded at points on the wreck earlier earmarked for comparative study. The engine was stropped (sandbagged and bound with wide webbing straps to spread the pressure) in preparation for its removal.

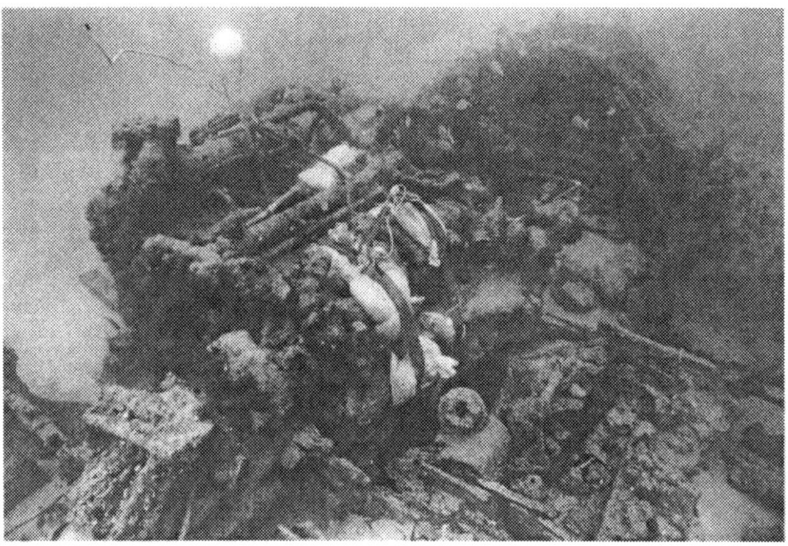

Figure 54. The engine ready for lifting. The strops and sandbags protecting the engine are evident. To the right of the engine are oil and other containers. The break between the flanges joining the engine and the thrust block is visible in the bottom left-hand corner (Pat Baker).

An external site grid (a reconstruction of the 1983 external grid) was set up and while these preparations were under way, underwater and abovewater video recording was used to augment the 2D and 3D color and black-and-white still photography that was common to underwater excavations. The excavation also commenced, but while this and the recording program were under way, a late-season cyclone was identified slowly progressing down the coast. As is the norm in Western Australia, it was preceded by a period of flat calms and as a result, it was decided to bring the lift forward one week ahead of time.

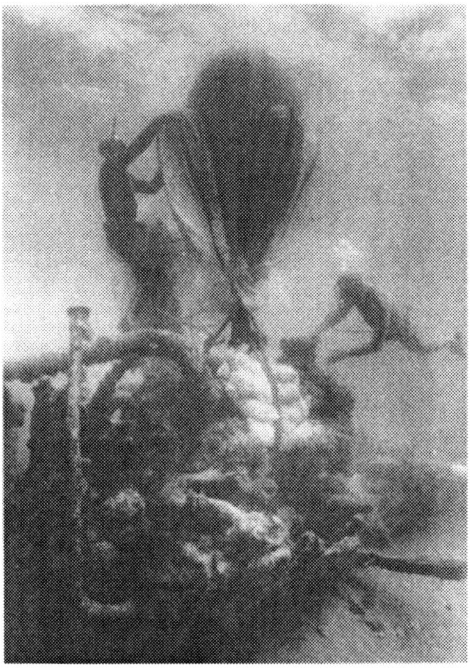

Figure 55. Riggers Geoff Kimpton and Brian Marfleet making final checks of the engine before lifting (Pat Baker).

7.8 The engine removed

Eventually 27 lifting devices (ranging from 1-tonne bags to inverted plastic tubs) were utilized and the engine was lifted and towed to the beach, where Kimpton, the chief rigger, had a prefabricated iron sled waiting. There the engine was run in toward the shore until it was located only a few centimeters above the sled. The sled was then lifted up to the engine using air containers and when the two met, they were firmly lashed together. The sled, with engine attached, was then towed ashore and secured on land ready for additional 2D and 3D recording, should it begin to disintegrate due to its removal from the sea and the increased effect of gravity.

Large iron objects, such as cannon and anchors, which need to be stored on-site awaiting transport, are usually covered in hessian and kept wet with sprinklers, or are preferably inundated in a 5% solution of sodium hydroxide (caustic soda) to keep them from further corrosion. Water was at a premium at Port Gregory, however, and the engine was a composite of cast and wrought iron, coppers and brasses. As a result it had been decided to use Erosel (a soil-wetting and stabilizing agent), combined with a solution of 1% sodium bicarbonate and 5% sodium carbonate soaked into hessian. Specialist on-site conservator Jon Carpenter had developed the technique a few years before to help preserve and pack fragile material raised from the wreck of an eighteenth-century wooden-hulled frigate (Carpenter, 1987).

Though completely covered with Erosel, wet hessian, black plastic, and ropes, a roster was set up to monitor and guard the engine. About an hour after midnight a 4WD car noisily pulled up and its totally inebriated occupants attempted to hitch up their vehicle to the sled and drive off with the engine and sled in tow (a totally impossible task). Angry words and threats were exchanged and there were offers to engage in fisticuffs, but nothing further transpired and the transgressors noisily departed. Next day they were to apologize for their actions, but at the same time they indicated that if they had known there was so much sellable scrap metal on the wreck they would have removed it long before and sold it to a scrap dealer. It was worth "a couple of cartons of beer at least" they said (personal communication to McCarthy April 18, 1985).

7.9 The excavation and recording of the stern

With the engine (XA 57) removed, the excavation of the stern aft of the boiler resumed. Standard techniques were used. For example, builders' levels to maintain the horizontal, plumb bobs to gauge vertical separation, together with two and three-tape systems for horizontal fixing.

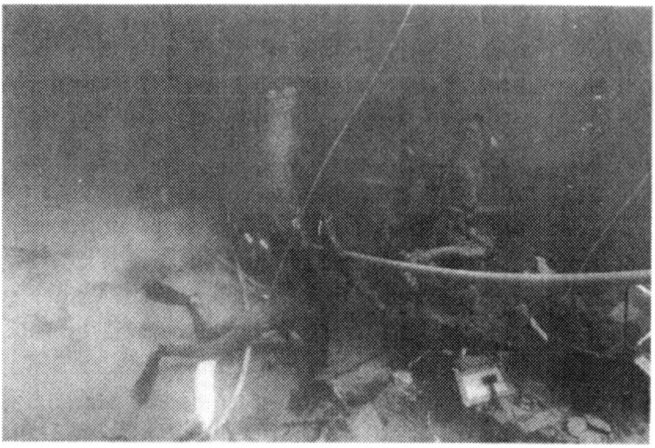

Figure 56a. View of the engine room taken from the aft port quarter (Jon Carpenter).

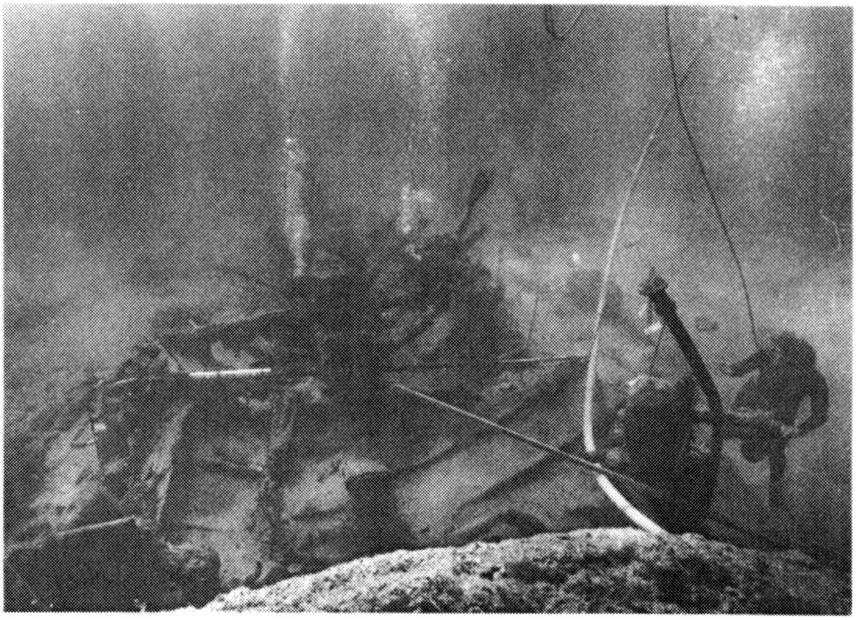

Figure 56b. View of the engine room and stern taken from the top of the boiler looking aft (Jon Carpenter).

Each evening the results were plotted on a 1:10 site plan while the conservators attended to the on-site conservation of the materials raised. Throughout the excavation, the stereoscopic system shown below was deployed in conjunction with video and the usual manual recording methods. It comprised a twin 15 mm underwater camera system mounted on a 1 meter-long bar and a double grid frame, which allows the diver to locate the cameras vertically above the target. When correctly in position above the dual grid frame, only one of the pair can be seen through the viewfinder. Intended as a three-dimensional supplement to manual recording, it provided a very useful cross-reference where checking or augmentation of the data in the laboratory was required (cf. Green, 1990).

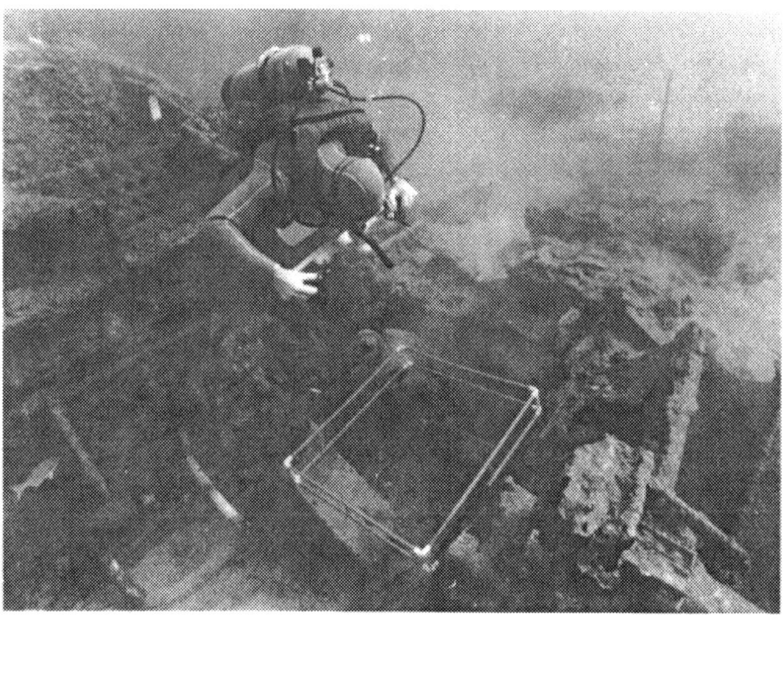

Figure 57. Photographer Pat Baker utilising his stereoscopic camera system and dual grid frame in the *Xantho* engine room (Nic Clarke).

Given our belief that the boiler was not cylindrical, a diving naval architect, Geoff Hewitt, spent considerable time re-examining and redrawing it in detail. His work (Figure 47) shows that though slightly elliptical it was once a cylindrical Scotch boiler, its anomalous shape being due to distortion on the seabed caused by the ingress of sand, corrosion, and other forces. Hewitt also recorded the piping arrangement between the boiler and the engine and produced a schematic analysis of the function of each component part.

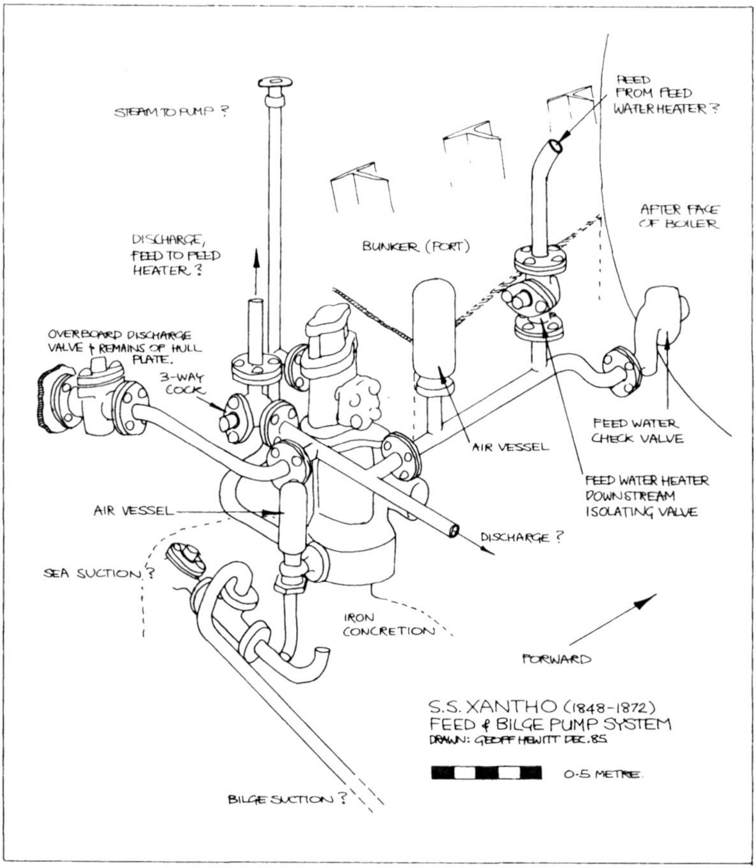

Figure 58. Geoff Hewitt's record of the piping and his analysis of its functions.

Though conjectural, the arrangement is based both on his knowledge of early marine engineering and his experience in the design of modern naval vessels. These representations indirectly attest to the need to have diving engineers along with the other specialists on sites of this complexity.

In this period John Riley, whose work on iron shipwrecks was discussed earlier, completed the isometric projection of the site, shown in Figure 45 and elsewhere throughout this work. It was also intended as a demonstration of rapid site recording method, and the practical demonstration was partly for the benefit of those seminar participants interested in the problem of recording and presenting sites with high relief in deep water, i.e., with short dive time. The isometric view was also of importance for those involved in the *Xantho* excavation, for where sites are large, extending beyond the limits of diver visibility, the isometric projection allows areas chosen for detailed excavation or examination to be conceptualized as part of the whole. They are also very useful as a briefing tool and are standard practice in iron and steamship archaeology (e.g., Murphy, 1987, 1998).

The excavation continued at the site until the cyclone crossed the coast, temporarily halting work. As numerous whalebones had been found on the wreck (including one firmly cemented to what proved to be the ship's three-bladed propeller) the lost days were spent in a successful search for a contemporary whaling station near the wreck (Gibbs, 1994). The presence of the whalebone low down on the propeller blade was a further indication of the periodical uncovering of the site (Figure 59).

7.10 The stern cut free

In completing the excavation, the stokehold and engine room were found lined with cement, as was often the case with iron ships. Beneath the cement original metal remained, though not of uniform thickness. At best, some plates were estimated to be 4 mm thick.

In the only attempt to excavate under the hull itself (for fear of causing damage), a trench was excavated down and under the starboard side of the hull just forward of the stern. This was conducted in order to ascertain the state of the ironwork and to gauge the depth to which the hull was buried.

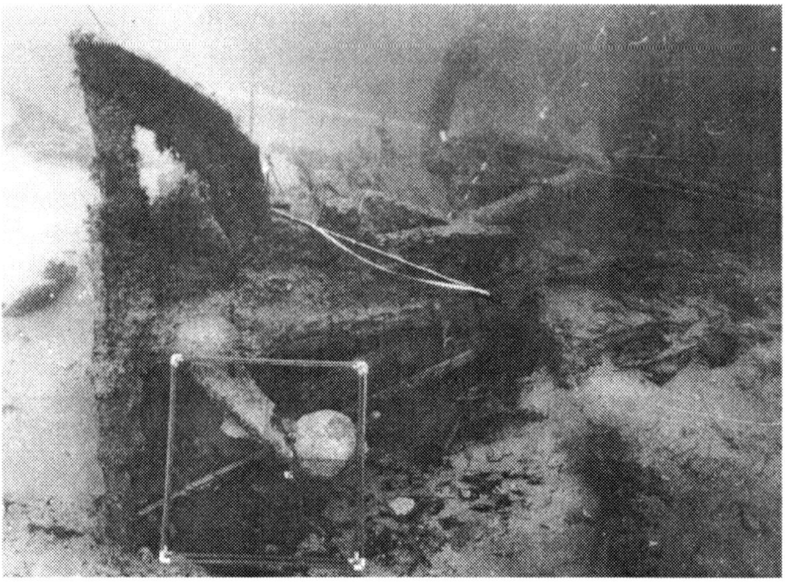

Figure 59. The stern of the ship and a whalebone found cemented to the tip of *Xantho*'s propeller (Pat Baker). The rudder is "hard a starboard" and the propeller is "right handed."

The hull on the starboard side of the stern, aft of the thrust block projected 700 mm above the seafloor. Below the seafloor a band of 200–400 mm of hull was covered by weed mat, leaving a further 800 mm of well-preserved hull in sand down to the keel. This indicated that a c. 2 meter high section of hull remained in the vicinity of the propeller aperture in the aft section of the ship on the starboard side.

After discussion with corrosion specialists and other conservators, it was decided to prepare this section for future recovery. The thermal lance was again used, producing a neat c.25 mm cut and after four days of solid work, the two meters of the stern aft of the thrust block (from rib 3 aft), were cut free from the rest of the hull. Even when cut, it would not move, however. Eventually two 75 mm diameter limestone pinnacles were found holding the keel fast to a hitherto undiscovered limestone reef underneath the sand and weed. These pinnacles were subsequently cut by hand, their size and strength providing another clue to site formation processes at the wreck. Though excavations were not conducted elsewhere under or alongside the hull, it was evident that some parts of it were most likely in contact with the limestone reef beneath the mobile sand. This would have added further stresses and would have contributed markedly to the breakup of the hull.

Kimpton then prepared the stern for a trial lift and air was slowly fed into the lift bags to the point of near-neutral buoyancy. The bags were then deflated and the stern section allowed to settle back on to the seabed. Anodes were applied, beginning prolonged *in situ* treatment for the time when appropriate facilities could be made available to receive the stern. A small section (XA 517) was recovered for study in the interim and an exhibition proposal, which included the stern and propeller, was developed and prepared for consideration.

The feed water heater, boiler valve, and what appeared to be an auxiliary pump on the post side of the engine room were subsequently covered with rocks to confuse and deter looters and work began on the cleaning up of the site and the finalization of recording. The expedition had deployed a total of 55 people for a total of 605 operator days and a total of 603 diving hours spent on the wreck over a one-month period from mid-April to mid-May 1985.

7.11 Subsequent on-site excavations

In February 1988 a visit was made to *Xantho* en route Kimpton and the author's excavation of the Dutch East India Company ship *Zuytdorp* (1702-1712). Conservators MacLeod and Carpenter joined us to assist with the conservation of a cannon and anchor earlier moved into deeper and safer water at the East-Indiaman (McCarthy, 1988c), and to monitor the anodes that were attached to *Xantho*'s stern in 1985. Conditions were suitable for diving, though the area aft of the boiler was covered in sand to just above the level of the top of the furnace doors. In contrast, the scour pit on the forepart of the starboard side of the boiler was deeper than previously seen, revealing more of the wooden bearers on which the boiler lay. This illustrated the mobile nature of the sand on which the *Xantho* lay.

During the inspection, the boiler relief valve and feed water heater were examined in more detail and a decision was made to recover them after the corrosion measurement study was completed. The finding of a spare eccentric strap under the feed water heater led to a surface search of the depression from which the heater came, and the slope to starboard, revealing a brass tap and some unidentified tools. These were recorded, catalogued, and recovered for conservation.

In March 1992, with the aim of further monitoring the site and removing an air pot and auxiliary pump assembly previously recorded, the *Xantho* was visited again en route to the *Zuytdorp* excavation. By then interest in the *Xantho* project had spread and we were joined by Paul Mardikian, a Paris-based diving conservator, working on materials from the iron-hulled steamers RMS *Titanic* and CSS *Alabama* (1861–1864) (Enault, 1988; Guérot, 1988).

The stern of *Xantho* was found covered by sand to the level of the counter and to the top of the fire boxes on the boiler. The bow area had less sand cover than normal, considerably exposing the starboard side from the bow aft to the boiler. Mardikian recorded electro potentials at all points previously measured, except those buried, and tested the anodes fixed in 1985. These were found to have worked well, being totally consumed and in need of replacement.

In March 1994, the wreck was again visited and was seen to be in what can now be referred to as its *summer configuration*, i.e., in comparison to the situation in winter, it was part-buried. Only the tip of the propeller was exposed, the boiler was visible aft to just below the furnace doors, and the area amidships was totally covered in sand. On the other hand, the stem was totally exposed, allowing it to be examined closely for the first time. The starboard forepeak was totally opened up by the current, revealing much more of the timber ceiling and other fittings than had been visible in earlier years. A rope fender was exposed just aft of the forepeak, for example. Being outside the areas relevant to the research strategy and not being in any apparent danger, it was reburied.

Again the anodes had to be uncovered by excavation, one lying alongside the starboard hull at the stern, the other inside the hull at a higher level. The upper one had been consumed and the lower was in original form. It had apparently ceased working soon after it was fitted, due to a break in its connections. This remaining anode was reconnected, a new one was fitted, and further potential measurements were taken.

The corrosion potential study was completed in this instance by diving conservator Jon Carpenter, adding to the data earlier obtained by North, MacLeod, and Mardikian since the first application of anodes to the site in 1983. These studies have proved the basis for a long-term comparative study on the use of anodes in the reduction of corrosion on underwater sites.

With respect to those anodes attached to the stern section, the surface pH there had reduced after 20 months of treatment. This produced an estimated 60-fold reduction in acidity. Thus the stern was undergoing conservation treatment *in situ.* (See MacLeod, 1986, 1987, 1990, 1992, 1995, 1998 for details.)

7.12 An unsuccessful attempt to examine nineteenth-century shipbuilding method

In returning to the hull for clues to its construction, a number of exposed areas were selected for examination in 1994. These were the stempost, a section of frame projecting on the port side of the boiler aft and a collapsed section of hull forward of the paddle-wheel sponson bearers. Earlier it was found that these were very heavily concreted and nothing of value was ascertained from a nondisturbance visual inspection. On the internal surface of the frames and hull there was not only concretion but also possibly cement, further obscuring the details of the original plating.

Particular attention was paid to a collapsed section of hull forward of the boiler near the starboard paddle-wheel sponson. It consisted of three frames and a 2-meter-square section of plate. Loose detritus was cleared using suction, hand fanning, and a small handheld water jet. Again the concretion/cement layer made the nondisturbance attempt to obtain accurate data by physical means or by photography nearly impossible. In attempting to examine the interior, gentle suction was applied to a break in the concretion, and clouds of black corrosion products streamed out into the water column. This left a hollow shell, but no information about the interior could be obtained *in situ.* As a result, the procedure was not continued. These experiences further highlighted the problem of making a nondisturbance record of features such as plate and frame scantlings on iron wrecks. This was correctly considered to be a fundamental problem confronting those involved with iron and steamship wrecks (see the discussion on page 132). Further, the interstices of the concretions also appeared too thin and fragile to produce *in situ* latex, polysulphide rubber, or similar type of casts traditionally produced in the case of heavily concreted objects (e.g., Murdock and Daley, 1982).

7.13 A clue to the rapidity of concretion formation

On this, the last dive on the wreck of the SS *Xantho* to date, a plastic-handled, steel-bladed diving knife was seen projecting from the sea floor 3 meters from the starboard paddle-wheel sponson. A roll of blue electrical tape carried on the handle identified it as that known to have been lost by the author in 1985.

It was very heavily concreted, far more than had been expected. The knife was recovered and preserved intact as a comparative specimen. It shows the results of a decade of concretion growth and corrosion on modern steel at the *Xantho* and provides an insight into what may have happened over quite a short period on the iron wreck itself. It also provides a useful example of the rapidity of the concretion process and is currently being used for such purposes in other studies (e.g., MacLeod, 1995).

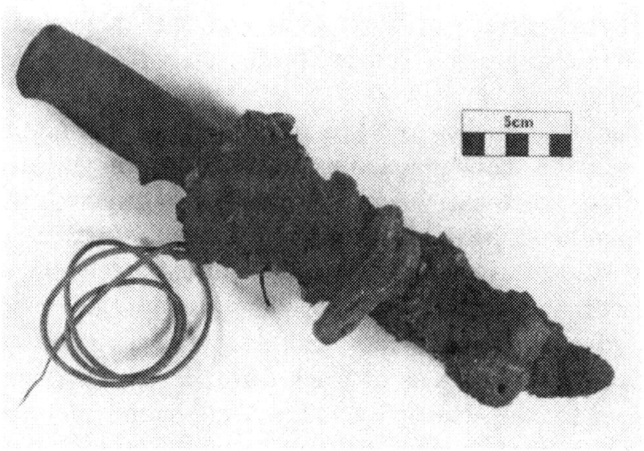

Figure 60. The author's knife after nearly a decade on the sea bed alongside the *Xantho* (Jon Carpenter).

7.14 Results of the 1984–1994 excavations

The interior of the stern section of the *Xantho* wreck up to and including the boiler face was totally excavated. The vessel's engine and most of its auxiliary machinery were raised, partly in the context of that excavation. The interior was recorded in three dimensions down to the cement lining in the stokehold. The difficulties experienced in taking accurate measurements of construction details through layers of concretion, and sometimes cement, on an iron hull (see discussion on page 132) were also acknowledged (cf. Lyon, 1984; Henderson, 1988).

A total of 178 artefacts (including the engine XA 57) were recovered, 14 before the engine was raised and the remainder subsequently. The assemblage was consistent with material expected in an engine room into which accommodation space had collapsed, e.g., concretions, glass and ceramic sherds, tools, oil containers, engine spares, boiler fittings, cup hooks, metal cabin fittings, fire bricks, door keys, a lamp glass, a box of matches, a salt cellar, brass taps, some timbers (possibly wooden cabin fittings), hessian fragments (from lead ore sacks), a small portion of the hull, and modern contamination. Small fragments of lead ore were found around the engine, indicating that some of the ore had been stored in the aft section of the ship between the boiler and the stern, most likely in the accommodation spaces and on the deck above the engine. This spatial patterning was consistent with the evidence given at the inquiry into the loss of the ship.

The small part of hull raised from the stern (XA 517) was analyzed by Maria Pitrun, a metallurgist with considerable experience in the analysis of shipwreck materials (MacLeod and Pitrun, 1986). It comprised a complete double riveted external butt-plate over the remains of two strakes of hull plating, with part of a frame attached (see also Figure 2). A section through one of the rivets attaching part of the butt-plate, the strakes of hull plating, and the frame appears in Figure 61 below.

In addressing the question whether corrosion or other associated problems had contributed to the sinking of the vessel, Pitrun advised firstly that though significant corrosion had occurred in the crevice-like spaces in between the sheets, she was unable to determine how much corrosion had occurred prior to the vessel's demise.

It was also acknowledged that the section raised for metallurgical analysis may have been added to *Xantho* during the 1871 refit. Altering the ship from paddle to screw propulsion required the fitting of a stern gland and a large propeller aperture and possibly a new section of hull aft.

While indicating that it was not inconsistent to blame extensive interseam and stress corrosion for the loss of the ship, Pitrun also concluded that the study was complicated by two major considerations. These were the periodic burial of the wreck and the excessive amounts of lead sulphide measured aft (downcurrent) of the lead ore in the *Xantho* cargo hold (see page 71).

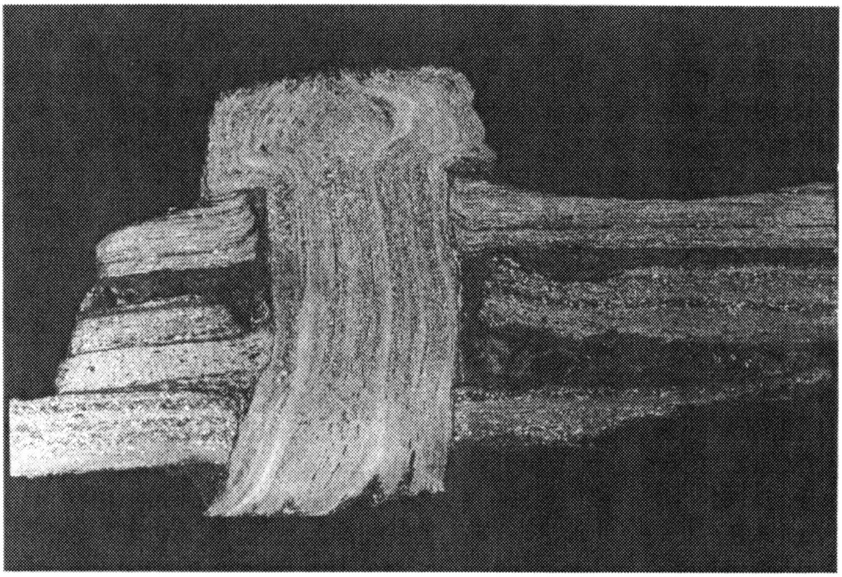

Figure 61. A magnified (X2) photographic section of a portion of XA 517 (Maria Pitrun). The section is through a rivet fixing a butt-plate (top layer), to strakes of plating, and a frame (bottom layers).

Chapter 8

Excavation in the laboratory

The deconcretion, disassembly, and conservation of the engine

8. DECONCRETION: EXCAVATION AND EXPERIMENTATION

The deconcreting and the disassembly of the *Xantho* engine was treated as a problem-oriented archaeological excavation addressing the hypothesis that Broadhurst was naïve and that the hybrid *Xantho* was poorly engineered and a poor choice of vessel for use in the north of Western Australia. As a result, archaeologists, conservators, corrosion scientists, technical staff and volunteers worked with the author as a team as they did in the water, analyzing and deconcreting the engine—as if it were an excavation.

Given that the *Indiana* engine mentioned earlier was recovered from a relatively benign environment in the Great Lakes, requiring little conservation treatment (Johnston and Robinson, 1993; Johnston, 1995), it could not be used for comparative purposes. Thus, the treatment of the *Xantho* engine also represented an experiment in materials conservation method. This required that traditional conservation and deconcretion method (cf. Green, 1990; Pearson, 1987) be modified and new approaches developed.

Because it was to be an experiment of relevance to the future development of iron and steamship archaeology and associated conservation science, the author also made it a requirement that both the successes and failures of this particular experiment be documented and promulgated. This was agreed to by all and the resultant techniques and understandings have provided those who follow a foundation on which to build (e.g., Carpenter, 1987; 1990; Garcia, 1996, MacLeod and Pitrun, 1986; MacLeod, North and Beegle, 1986; MacLeod, 1989a; 1989b; 1990; 1992, 1998, 1999; North and MacLeod, 1987; North 1997; Pénnec, 1990). A précis of the results and the methods used is presented in what follows partly also to illustrate that the excavation of the engine was both allied to, and dependent on, conservation method.

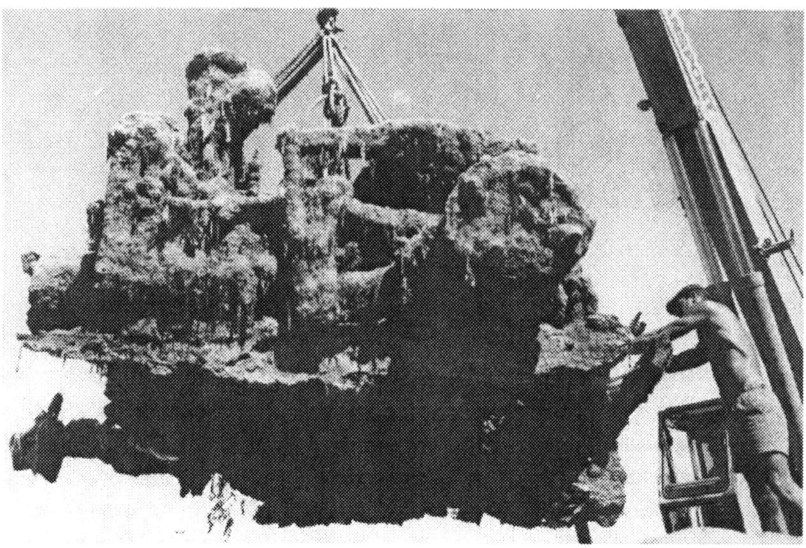

Figure 62. Geoff Kimpton with the engine before placement in the treatment tank at Fremantle. The flange (to which was joined a corresponding flange on the shaft) is on the right, radius link center and valve chest and trunks left. The remains of the engine bed are visible beneath (Pat Baker).

8.1 The removal of chlorides explained

When cast and wrought-iron objects are submerged in a saline environment they corrode at a rate dependent on a wide range of factors; notably water movement, biological activity, temperature, dissolved oxygen, the presence of dissimilar metals, and salinity. In essence, the corrosion results in the inward diffusion of chloride ions through the concretion into the corroding metal and the outward diffusion of metal ions back out through the concretion to the sea. This results in there being an inner core of unaffected original metal and (especially in the case of cast iron) a graphitized outer layer. This layer contains little or no iron and a heavy concentration of chlorides (salts).

The treatment process is designed to speed up the diffusion of chloride ions from the graphitized layer into the treatment solution. With large objects the method used to achieve this first involves the stabilization of the iron, using a solution of alkalis (5% sodium sesquicarbonate and 5% sodium carbonate, or 2% sodium hydroxide) in water. It is a lengthy process part-pioneered and refined by Neil North and his predecessor Colin Pearson (North and MacLeod, 1987). This entails making the iron object a cathode within a tank of 2% sodium hydroxide or 5% sodium sesquicarbonate. The anodes are sheets of mild steel and electrolysis occurs when a current is passed through the metals. By these means, salt is released from newly deconcreted iron. There is the risk of losing surface detail if the current is too high, however. Excessively high currents cause an evolution of gases that can force the fragile graphitized surfaces from the remaining metal (cf. Oddy, 1975).

Where possible, hydrogen reduction is performed. This procedure involves the heating of corroded iron to 400°C in a flow of hydrogen gas; iron oxides and sulphides are reduced, giving off steam, hydrogen sulphide, and hydrogen chloride gas. Hydrogen furnaces are normally quite small and the metallographic structure of the uncorroded iron can be altered at temperatures above 450°C.

8.2 Deconcretion of the outer surfaces

The deconcretion process not only reveals the original structure for archaeological and technical recording but also assists in the conservation process by allowing chlorides to escape from the surface of the object. As a result, the chloride levels in treatment solution in which a recently deconcreted object is placed will rise. When these rise to a point where they match that of the chlorides in the deconcreted iron, the diffusion of the salts across the surface of the iron and into the solution stops. When this occurs, or when it is slowed to a marked extent, the tank is emptied and filled with a new solution.

Following the arrival of the *Xantho* engine at the Western Australian Maritime Museum in late April 1985 it was weighed at 7.4 tonnes, well within the range of the engineer's estimate (see page 126–127). It was then lowered into a treatment tank containing a solution of 40 kilograms of sodium bicarbonate and 60 kilograms of sodium hydroxide, to inhibit further corrosion.

In May the tank was drained and the first deconcretion session began. A water spray was used to keep the engine from drying out and this was applied as required at intervals during the day and throughout the night. Not having been attempted on such a large and complex artefact before, considerable experimentation was required.

First "foreign" objects, such as two iron-haired cylindrical boiler tube brushes (XA 222), were examined and chipped free. These are visible in Figure 29, but were initially a mystery and required X-ray analysis to confirm their identification, as did a pair of what appeared to be gasket scissors. Concreted rope, tools, copper pipes, lead ore, boiler sight glass fragments, screws, and other smaller objects were also found among concretions on the engine. These had apparently fallen onto the engine from above or from the port side of the engine room.

Removing the numerous brass lubricators, tallow pots, cocks, nuts and bolts screwed onto the engine presented little difficulty. The layer of concretion on them was only a few millimeters thick and their threads were in pristine condition. In contrast, the brass maker's nameplate was found under a thick layer of concretion on the iron valve chest. It was further deconcreted and found to be in perfect condition and bearing the legend J. PENN & SON ENGINEERS GREENWICH (Figure 12). The nameplate provides an instance where the written record was corroborated and in part corrected by the physical evidence. While the 1871 register recorded the correct details, one contemporary newspaper referred to the engines as built by "Payne, the government machinist," for example (*Herald*, January 25, 1873).

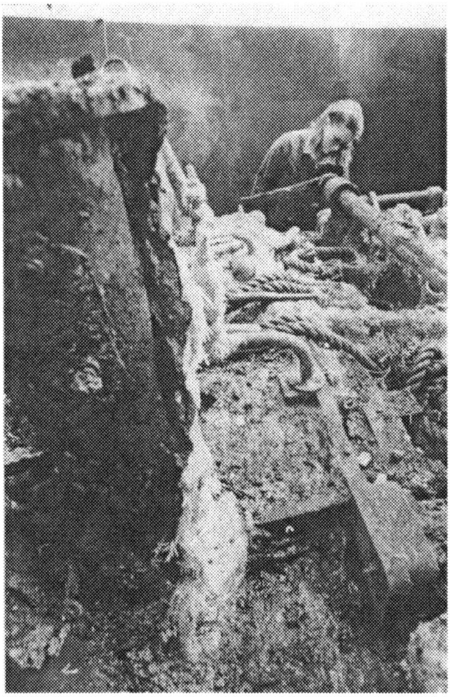

Figure 63. Deconcretion in progress. In the foreground is the upper part of the engine exhausts, showing the thickness of the concretion layer. In front of conservator Michelle Berry is the wrought-iron weigh shaft (Jon Carpenter).

The deconcreted copper and copper alloy items were treated using 5% citric acid and thiourea in the deconcretion phases and then desalinated in a solution of sodium dithionite and sodium hydroxide in order to remove the chlorides. After cleaning they were prepared for storage or exhibition.

Brasses, copper, wrought and cast irons are typically found on marine engines and wood is often found as lagging (or insulation) around the cylinders. A surprise find was an engineer's footrest; a shaped timber that was fixed onto the engine bed and ran parallel to the crankshaft. This prevented the engineer's foot from slipping as he attended to the engine and it was located so close to the crankshaft that it had to be scalloped out to avoid contact with the big-ends of the connecting rods. Being wood, it was removed for separate treatment (XA 273).

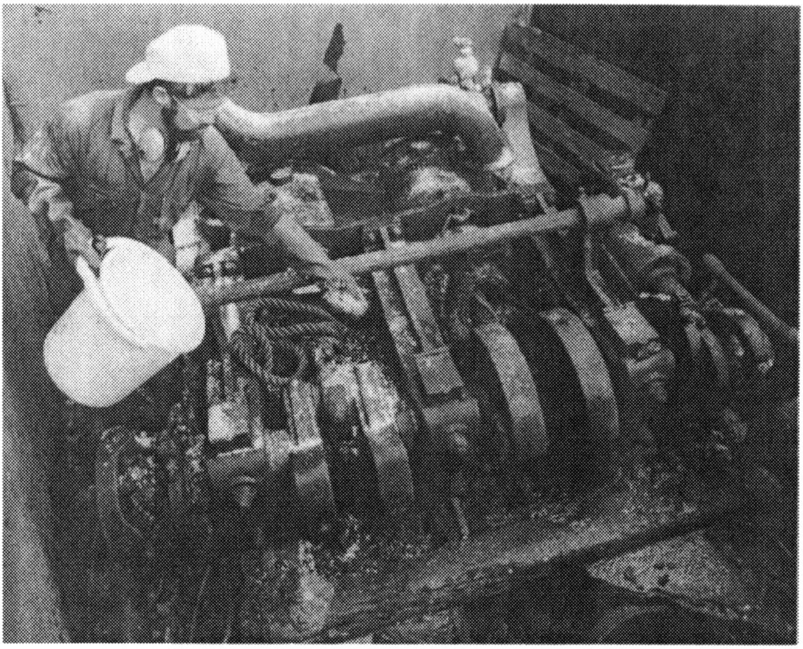

Figure 64. A view and closeup of the partly deconcreted engine showing the wooden footrest just below the counterweights in the foreground (Jon Carpenter). Ian MacLeod is sprinkling caustic soda before the water sprinkler is turned on overnight.

The wrought and cast iron structures, such as the cylinders, cranks, valve chests, pipes, mainframe, and the trunks were far more difficult to treat. These were covered totally with a layer of concretion ranging from 25–50 mm in thickness. Often the layer was thicker than the object buried within. The principal method initially used to remove the concretion was hammering and chipping in the traditional "percussive" mode (Pearson, 1987; Carpenter, 1990). This involved the use of tools such as hammers, geopicks, and chisels, the principle being that it is possible to separate the concretion and artefact at their junction by applying compressive or shearing forces. A trial application of what appeared to be more "acceptable" techniques, such as chemicals and ultrasound methods, had earlier proved unsuccessful given the thickness of the concretions. It was frustrating, backbreaking, dirty, and difficult work.

Ninety-two kilograms of concretion were removed in the first week. Over the weekend the engine was kept wet with sprinklers and continually monitored. On the first day of the following week a total of 350 kilograms of concretion and corrosion products were removed, revealing the upper surfaces of the trunks, the pumps, the tops of the cylinders, and cranks. As is the norm in excavation, progress was recorded manually, photographically, and by video; engine parts were catalogued as they were removed and daybooks were kept. As each engine component was sent for treatment it was labeled and tracked through the conservation process for storage or exhibition for eventual reassembly.

After this first week of experimentation a system was developed. Work progressed from 9 A.M. to 5 P.M., averaging six operators in the tank at any one time. In acknowledgment of the fact that the deconcretion process was both an experiment and an excavation in its own right (McCarthy, 1989), this included Ian MacLeod (as head conservator) and author (as the archaeologist) full-time. This strategy also reflected the understanding that the deconcretion process and the conservation of the engine was an experiment of the archaeologist's (and not the conservator's) making—i.e., the archaeologist who raised the object bore the ultimate responsibility for any failure.

Finally, it was also accepted that deconcretion of any complex object will inevitably cause damage:

> The concretion that is formed both in aerobic or anaerobic conditions can be exceedingly tough so that when conservators and archaeologists attempt to remove the marine growth, a significant amount of damage occurs, often with the loss of archaeological information (MacLeod, 1987:49).

After two weeks, 1,250 kilograms of concretion had been cleared from the upper surfaces of the engine. With the operators exhausted and with other pressing duties, the excavation was halted and the tank was filled with a new solution.

The tank was again drained as the chloride levels in the treatment solutions rose and during a one-week deconcreting session in July a further 555 kilograms of concretion were removed. The drop in weight of concretion removed by over 50% is an indication that, after the initial period of rapid deconcretion of large objects and their surfaces, attention was diverted to finer work.

As an example of the process, at the end of October 1985, the salt levels in the tank were measured at 727 parts per million (ppm). The tap water originally used to make the solution had 105 ppm salt content, indicating that a sixfold increase in salt content had occurred since the tank was filled after the last deconcreting session. As little change occurred in the salt level over the following week, the time had come to again drain the tank and so the excavation recommenced.

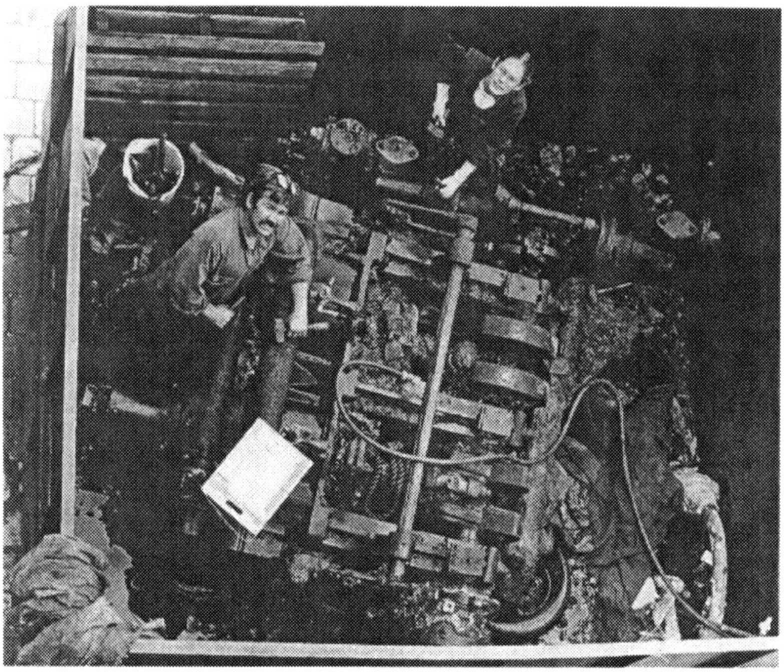

Figure 65. A view into the deconcretion tank. Conservator Nancy Mills-Reid is working on the port pump. Connecting it to the starboard pump is an engine-driven "Scotch Yoke". The author is standing between the trunks. Note the tools being used.

As indicated by the underwater surveys, corrosion rates over the engine were not uniform. In some places iron nuts and bolts had totally disintegrated or were present merely as hollow shells; in others they were intact. The fore wrought-iron radius link, for example, was found to be a hollow concretion due to it being preferentially corroded by the cast iron. Its concretions were kept for molding as casts for reconstruction in the future. The corresponding link on the aft (downcurrent) side of the engine was in far better condition, however. The wrought-iron weigh shaft, which joined the two, was similarly badly affected and it soon became evident that all of the wrought iron on the engine, including the heavy engine bearers, had suffered in a similar fashion. This contrasted with the brasswork and copper, which, as one would expect from their place in the electrolytic table, were uniformly in good condition.

Though the core deconcretion team remained the same throughout, it was joined at odd times throughout the next 10 years by staff, volunteers, and visiting conservators, some on internship from a number of countries throughout the world.

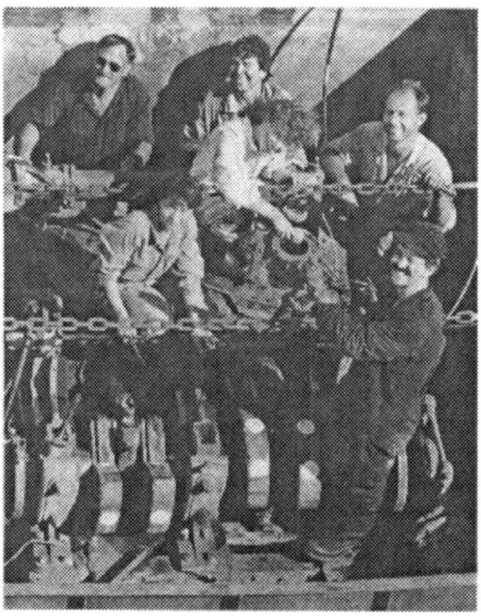

Figure 66. A deconcretion team in the tank. From the left—Dick Garcia, (Western Australian Museum), Kolam Gaspar (Brunei), Alexandra Elliott (Canada), the author, Paul Mardikian (France), Ian MacLeod (Western Australian Museum). The wooden footrest was bolted to the bearer under MacLeod's left foot. The chains are stays for the treatment tank (Jon Carpenter).

In looking for further evidence as to how the engine and the *Xantho* were operated, evidence of hasty repair was sometimes found. What was described in the artefact book as a "rough rubber gasket with wire" (XA 288) was in fact the product of a very crude attempt to stem leakage of steam between the inlet valve on the fore valve chest and the valve chest itself. Further, the next deconcretion session, which began in early June 1986, resulted in our steam engineer Noel Miller bringing the direction of engine rotation into focus.

8.3 Engineering anomalies identified

While recording an oil cup on the crankshaft, Miller noted that it was designed to take lubricant on each rotation from an oil-impregnated wick suspended above the crankshaft. This in itself was not unusual. The surprising feature from Miller's perspective was that the cup was open to the starboard side of the ship. This indicated that, when viewed from astern, the crankshaft and hence the propeller rotated in a clockwise direction and the top of the propeller and crank moved from left to right–i.e., the propeller and the engine had a "right-handed" rotation (*Xantho Engine Deconcretion Book*, May 23,1991).

Though the fact that the propeller was "right-handed" had been noted during the site inspections, its significance was not understood until Miller applied his knowledge of marine engineering and its history. John Penn had designed all his trunk engines fitted to the port side of the ship (like *Xantho)* to rotate anticlockwise, in order to counteract the force of gravity on the cylinders and thus reduce their wear (Seaton, 1911). Even by the 1920s, teachers of marine engineering felt a need to describe the type and to explain this particular feature, as the following excerpt shows:

> A trunk engine has no piston rod, but simply a connecting rod extending between the crankpin and the piston. The trunk passes through both ends of the cylinder, and is bolted by its flange to the piston. For a right-handed propeller this type of engine is placed on the starboard side of the engineroom, so that when going ahead the stress will be thrown on the top side of the trunk and piston; for a left-handed propeller the engine would be placed on the port side to obtain the same results (Sothern, 1923:299).

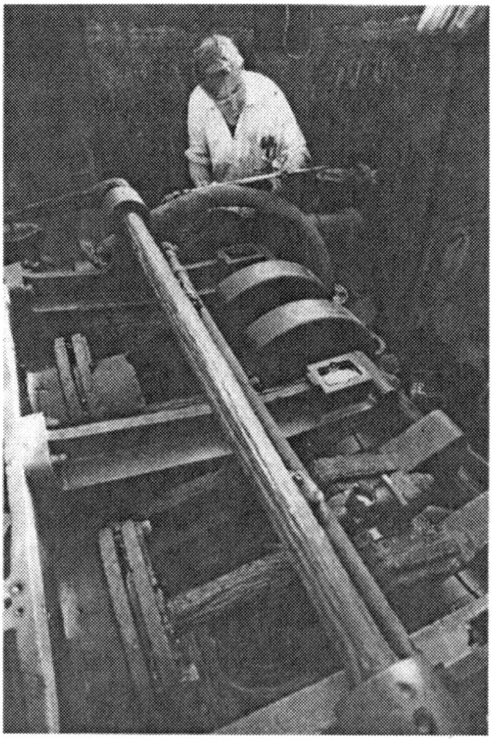

Figure 67. Noel Miller recording the upper surfaces of the engine. The oil cup is in the right foreground (Pat Baker).

The *Xantho* engine was rotating in the opposite direction to that which its maker had designed. This was a most anomalous find, as the direction of rotation on the *Xantho* engine served not to lift the piston up from the cylinder floor on each rotation, as Penn had designed in order to reduce wear (cf. Jamieson, 1897). The direction actually served to force it downward, adding to the gravitational force and thus increasing engine wear markedly. This made complete engine failure only a matter of time, raising further questions about Broadhurst's decision to purchase the *Xantho*. Did the problem exist when he purchased the ship? If not, when had the propeller been fitted? Was he even aware of it? Why did *Xantho* not carry a spare, when its successor *Georgette* carried a spare propeller for use on a run far closer to engineering facilities? Whatever the results, for a vessel to have an engine running backward to proceed forward added further to the belief that Broadhurst or his engineering advisers were in some way deficient.

8.4 Further deconcretion

In late October 1986, the tank was again drained for another week of work and, in view of the hardness of the remaining concretion, an air-driven needle gun, an air-driven chisel, and an air-driven drill were tested, albeit with some reluctance. The last two were utilized to effect and though the potential for damage was again noted, these tools were considered suitable due to their efficiency in cramped spaces.

Eventually, few artefacts remained attached to the engine structure, leaving mainly nuts, bolts, and a few remaining oil cups and other brass-work pieces to be removed. It was slow work, however, and in 1986 only nine objects ranging from an iron bolt to rubber and wood fragments were removed compared to the total of 345 recovered up to that time.

The next stage of deconcretion was conducted over two and a half weeks in late September to early October 1987. It involved firstly the raising and reshoring of the engine with timber to allow work to begin underneath. For this purpose Kimpton had designed and constructed a heavy metal frame on which to rest and raise the engine. Then 300 kilograms of concretions varying from 30–50 mm thick were removed from the underside, the relatively large amount reflecting the fact that work was being performed for the first time in this previously inaccessible area.

The dismantling of the inlet and exhaust pipes and other piping that once linked the boiler to the engine and it to the pumps also began, as did the deconcretion of the area inside the trunks. This involved the use of percussion, air tools and masonry drills.

In this phase 547 kilograms of concretion were removed, making a total 2469 kilograms removed up to that point.

Finer deconcreting amongst the interstices of the engine resulted in the removal of a number of brass fittings such as the trunk gland tightening nuts (XA 349–362). A kitchen knife (XA 366) was also recovered from one of the concretions, again providing some insight into the disintegration of the accommodation spaces above the engine room. Work also continued on the deconcretion of the inside of the trunks. At the completion of this phase the tank was again filled and the electrolysis continued.

8.5 Unexpected conservation problems

There were also a number of unexpected problems in this stage. In August 1988, for example, cast iron fragments from the edges of 14 discrete surfaces on the pumps were found to have detached between treatments. Apart from his own assessment, Ian MacLeod facilitated a series of analyses. One was conducted by a member of his visiting staff Stephane Pénnec, a conservator later involved with the treatment of RMS *Titanic* material, and another by Neil North, who was then in private practice. In essence it appears that the "exfoliation" had two interrelated causes. The first was the inability to maintain a uniform field of current throughout an angular object without the surface symmetry of the iron cannons usually treated. In this case the sharp right angle edges on the pumps and other fittings presented a relatively new problem to the conservators. In these instances the graphitized layer at the corners of an object becomes very vulnerable. The other cause was the inadvertent application of more current than was necessary. The subsequent excessive evolution of gas caused the fragile metal to spall off, especially at corners where it was most at risk from external forces (Pénnec, 1990).

8.6 Further evidence of abandonment behaviour

The tank was drained and solutions were changed in December 1990. In May 1991 it was drained again and a further week of extensive deconcretion was carried out. This proved most successful, and it included much fine work and close recording by Miller with a view to producing engineer's drawings of the engine. He also noted that the engine was in midgear, i.e., it was stationary and was not driving the propeller either forward or in reverse.

A slowly sinking vessel can be expected to have its engines in midgear, in order to facilitate an orderly abandonment by reducing the danger from a rotating propeller and in order to bring the vessel to a halt so that lifeboats can be launched. Steam is still needed to drive auxiliary machinery, however. Where a vessel is hard aground, and cannot be got off, the engines are also expected to be in midgear for the same reason. Only where a ship is rapidly sinking or where abandonment is abrupt, or panic driven, would the engines be left in forward or reverse gear by crew fearful of being trapped below. The engine room staff and stokers are among the most vulnerable in any steamship wreck and evidence of their abandonment behavior and by inference clues to the circumstances of the loss of any steamer are expected to be found in the engine room. Where there is an abrupt abandonment of the lower decks, the stokers would hurriedly abandon their posts, sometimes leaving the fire doors open, for example. In the *Xantho* case they are found shut (Figure 47).

8.7 The externals totally deconcreted.

Further deconcretion work was conducted in March and October 1992 and, by the end of seven years, the engine lay almost completely deconcreted with all but the internal parts of the trunks, cylinders and valve chests as clean as the day the *Xantho* was lost. The visible parts were stable and looked as if all they needed was a coating of wax and the engine could go onto the exhibition floor. In the process, 2544 kilograms of concretions had been removed from its surfaces and 48 kilograms of chloride ions had been released from within the actual metal of the engine itself (MacLeod, 1992a).

On the negative side, the area inside the trunks had proved too difficult and work in these spaces had been halted. The thin (10–25 mm thick) cast iron piping and the thin valve chest walls and cover plates also had little original metal left in them. As a result, their removal and work on the internal cavities of the engine were not envisaged.

An examination of the interior of the engine and the deconcretion of the trunks was necessary in order to complete the archaeological and technical analysis of the engine, however.

Earlier, there had been an unexpected effusion of gases from a number of internal spaces when drainage taps and cocks were removed. This indicated that the opened sections were dry. An endoscope was applied to these spaces, revealing some areas that had not been flooded when the ship went down. Given that entry into these internal spaces for the purposes of deconcretion was not considered possible, it was decided that they would be filled with corrosion-resistant vapor phase inhibitors to prevent the engine from corroding internally. The author's daybook of the time reads thus:

> Both the valve chests and the cylinders were examined and proved remarkably free of concretion with surface corrosion only and very sharp edges to all surfaces. This is a wonderful boost to us all and the light at the end of the tunnel now looms large: with vapor phase inhibitors it is clear that the cylinders and valve chests will not become a source of continued corrosion in the future and we can concentrate instead on existing concretion in the trunks themselves (*Xantho Engine Deconcretion Book*, May 23, 1991).

It was also envisaged that when rendered internally and externally stable, the engine would be coated in microcrystalline wax and placed on exhibition. Further recording work would be conducted in the public gallery, including the completion of the engineering drawings and the recording of the various oil cups, nuts, bolts, and studs and their threads.

Though it was still to be completely deconcreted, Miller also expressed surprise at the rudimentary construction of the engine bed. On the basis of his understanding of marine engineering, he had expected it to be far stronger. His impressions about the box frame on which the engine lay in the *Xantho* hull were that it was built with little real appreciation of the stresses and strains that would occur on the open sea.

8.8 The engine model

The philosophy behind the production of working models or replicas of excavated ships and boats, their construction, and their use in testing has been covered at length by a number of authors (e.g., Claasen, 1983).

In the *Xantho* case, excavation of the hull for the purposes of obtaining the ship's lines was not an option due to its fragility. As a result, the production of lines for the underbody of the ship and a resulting hull model or replica capable of being used for tank tests and other experimental purposes was not attempted.

Once it was decided to raise the *Xantho* engine, the production of a working scale replica, based on engineering drawings, became desirable.

It was to be the first model of a marine steam engine recovered from the sea. Miller, an accomplished model maker, had spent a great deal of time measuring and recording each section of the engine as it emerged from its layer of concretion, partly with that aim in mind. He became gravely ill, however and the task was taken up by his colleague, steam-engine model maker C. E. (Bob) Burgess. After a number of years and thousands of working hours he completed both a wooden mock-up and a 1:6 working scale model of the engine, which was presented to the Museum in May 1991 (Burgess and McCarthy, 1994). A suitable handover ceremony was held and in turn Burgess was presented with one of two replicas of the engine nameplate, the other going to the HMS *Warrior* in England (Figure 12). Soon after this ceremony both Marjorie Darling and Noel Miller died.

By reproducing each external part faithfully according to Miller's drawings, Burgess has enabled us to understand the engine as a unit and to visualize how it appeared statically and in motion.

What Miller and Burgess could not do, however, was reproduce the internal workings. These were developed from their understandings of nineteenth-century valve and piston systems. Up until then the only glimpse of what was inside the engine was through the lens of an endoscope, and that had proved inconclusive, except to confirm that some of the interstices were still in a very good condition.

As Burgess assembled the model progress was recorded until finally it was ready to receive compressed air, an accepted alternative to steam in the engine-model world. Not having dealt with a trunk engine before, Burgess was not sure that the engine would work. It did, however, and from a technical point of view the model has been an undeniable success, proving that not only was the gunboat type of trunk engine functional but it possessed a number of engineering advantages.

The model also demonstrated that the *Xantho*'s engineer was in grave danger of being crushed between the rotating crankshaft and the wooden rest. With the engine rotating in a contrary fashion to that designed by Penn and Son, the counterweights and crankshaft were moving down toward the engineer's foot-rest and not up and away from it, as planned. Though the footrest has been removed for treatment, these dangers are clearly illustrated in Figure 63 above, where a counterweight is poised just above Ian MacLeod's foot, showing that there was no margin for error in the circumstances described above. Here was another clear example of poor engineering practice on *Xantho*.

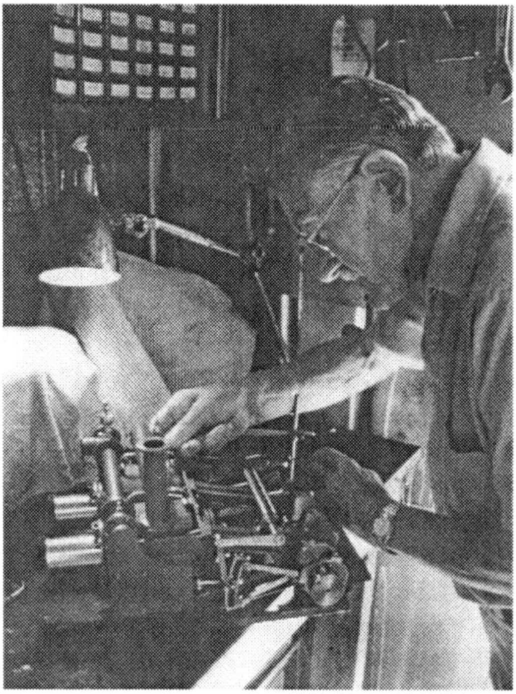

Figure 68. Model engineer Bob Burgess with the model in his workshop (Jon Carpenter).

The working model now has pride of place in the *Xantho* gallery and can be operated by the public, providing people with some insight into early marine engineering. The model has also enabled the team to examine the working engine over what is now an extended period of time. Insights were also gained into its compact and easily accessed nature, its change from forward to reverse or midgear, how it was supported on the engine bed and compactly housed within the vessel. Its simple and easily maintained nature also became evident.

8.9 Engine markings: Broadhurst reassessed

The surfaces of the cast iron structures on the deconcreted engine, though extensively graphitized, were uniformly intact. A series of numbers and letters were visible on the valve chests, the end of the trunks, and the brass-work, and eventually a pattern was seen to emerge. The majority comprised the letters 58 30 F or 58 30 A, indicating that each unit carrying part number 58 30 F was to be attached to the fore cylinder and that those with the number 58 30 A were to be attached to the aft cylinder. Occasionally the numbers 1–4 and position fixing dots were located. On the eccentric straps, of which there were two on each side of the engine, the figures FB and FT (fore bottom and fore top) and correspondingly AB and AT (aft bottom and aft top) appeared. More inscriptions were found as the "excavation" continued.

These markings attested to the possibility that the engine was number 58:30 and that it was capable of being readily assembled and disassembled, with its parts suitably identified and marked to ensure their correct order and place of application. This led to the belief that this was the 30th engine produced by Penn and Son in 1858. As the engine is recorded in the *Xantho* register as having been built in 1861, the question was put whether it had been stored for the intervening three years before being fully assembled for its fitting into a gunboat.

The matter was not satisfactorily resolved, however, for there was an exception to this uniformity in the form of four valves that were identified by Miller as "indicator cocks" (XA 311–314). These connect an "indicator gauge" to the cylinder and this is used to assess the performance of an engine by measuring the pressure of steam throughout its entire cycle. The indicator gauge also produces the data in diagrammatic form (an *indicator diagram*). Three of the four indicator cocks appear related, bearing the inscriptions 44 3F 30 (MACK), 44 T 30, and 44 A 30. The fourth is marked 64 3A 30. They all carry the government-issue broad arrow, which is not found on any other part of the engine. The inscription MACK appears to be a maker's name and again this does not appear on any other part of the engine.

Given that there was a need to mark them all with the broad arrow (when it does not appear on any other part of the engine), they may have been easily portable and may not have been part of the original machine. These "cocks" could have come supplied with the indicator gauge itself, however, and as engineers tend to keep tools of their own, especially precision instruments, these may have been issued as part of their professional "kit." Despite this, they do cast doubt on the notion that the 58 in the inscription 58:30 refers to the year the engine was constructed.

Eventually, deconcretion of the spare connecting rod, which was found in the engine room of the *Xantho*, revealed the legend SPARE

FITTED 58 30; the first indisputable proof of the existence of interchangeable parts for the engine. This composite (iron and brass) artefact, which is visible under the thrust block in Figure 45, was sent to the hydrogen furnace for treatment. This proved so effective that it was placed on exhibition in the Maritime Museum's front gallery soon after the application of a layer of microcrystalline wax. This is a standard procedure designed to protect conserved objects from the atmosphere and from the effects of handling.

More importantly the spare connecting rod, when combined with the ubiquitous markings 58:30 F and 58:30 A, indicated that the engine was designed to be easily assembled and that it was indeed of a mass-produced type that would potentially allow the use of interchangeable parts from other similar engines.

Broadhurst's decision to purchase this particular ship–fitted with a very simple engine that had easily fitted spares and interchangeable parts–came to assume another dimension. Was his decision based on logic, after all? Was it wrong to hypothesize that he was naïve and had been duped through his lack of understanding of marine engineering?

On reflection, it became apparent that an entrepreneur with limited funds, planning to operate a ship in a remote part of the world, in confined and shallow harbors, with poor coal supplies, few engineering facilities, and little freshwater might have preferred a vessel that was cheap and multipurpose. He may have consciously sought a ship fitted with compact, easily repaired machinery. Was the decision to purchase the *Xantho* a reasonable one after all?

Further, though it does not appear mentioned in any of the contemporary accounts, or registers and though, on paper it had a powerful engine, Broadhurst may have actually used *Xantho* as an "auxiliary steamer"–i.e., as a vessel that used its engine as a secondary source of propulsion. Given the operating parameters, was this more appropriate than utilizing it as a steamship proper, with auxiliary sails? Using an engine as the secondary, rather than the primary, source of power obviated the need for regular coal supplies, engineering facilities, and proper boiler feed water. To some extent it also mitigated against the thermal inefficiencies of the trunk engine type.

Broadhurst may not have wanted to advertise the fact that the ship was being used as an "auxiliary steamer" in order to make it appear more suitable for use as a coastal "tramp steamer" and thereby pick up more cargoes and passengers when not needed for pearling. Whatever the explanation, a new set of possibilities had been established.

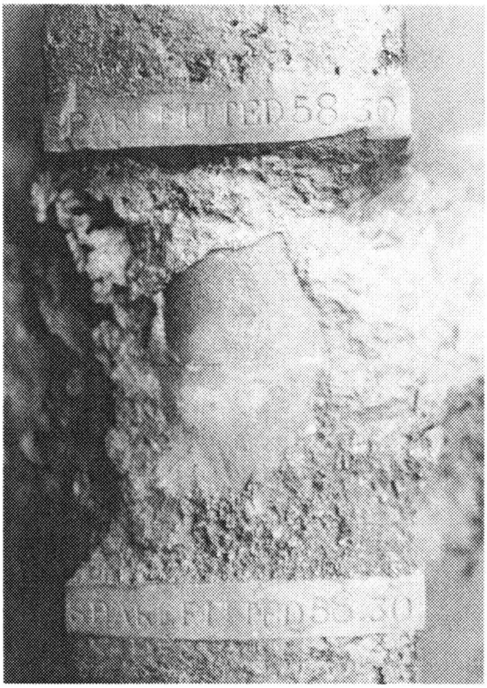

Figure 69. An end of the spare connecting rod showing the markings, SPARE FITTED 58.30. The connecting rod is shown *in situ* beneath the thrust block and flanges in Figure 45 (Jon Carpenter).

8.10 The disassembly of the engine

Another change in direction for the *Xantho* project came with the involvement of conservator Richard (Dick) Garcia. He had found over many years of conservation and restoration practice that the internal parts of artillery pieces and munitions could be freed and the corrosion products loosened using direct heat on their corroded outer surfaces. This *direct flame method* causes differential expansion of the corrosion layer and the underlying metal in addition to the production of steam at the metal/corrosion product interface. If expertly applied, so that the inner layer does not get unduly heated, the method causes the corrosion products to break their grip.

Having successfully applied the method to the deconcretion of a pair of previously untreatable fragile iron trypots recovered from the nineteenth-century wooden-hulled whaler *Lively* (c. 1815) (Carpenter, 1990), Garcia suggested the method could be applied to the *Xantho.* Initially the suggestion was met with considerable concern and scepticism on the author's part.

8.10.1 The direct flame method applied to the *Xantho*

As a comparative test both percussive and direct flame techniques were applied to a very fragile container that had been recovered from *Xantho* in 1985 and had been considered impossible to treat. Having proved successful, the test was then applied to the feed water heater (which was recovered in 1988) and the auxiliary pump (which was recovered from the port side of the stokehold in 1992). The percussive technique resulted in the usual surface chips, while the flame method showed little visible damage at all. Eventually it was concluded that the method was superior to percussive methods in confined spaces or where the remaining concretions were proving almost impossible to remove (*Xantho Engine Deconcreting Book*, March,1992). In March 1992, Garcia was authorized to conduct a test inside the trunks, where deconcretion had earlier proved impossible. Assisted by MacLeod and the author, the work proceeded successfully and a total of 75 kilograms of otherwise immovable concretion was taken from within these confined spaces.

8.11 A concretion formation model proposed

Of importance to the study of site formation processes on iron and steamship sites is the analysis of the concretions found within the trunks. These were not uniform in composition, thickness, or hardness, posing questions about their formation and the ramifications of this phenomenon for the materials preserved inside. Paul Mardikian, the French conservator who earlier had joined us on-site, was able to provide a preliminary model for the formation of concretions and corrosion-products within interstices (like the trunks) that are rapidly sealed. (His preliminary analysis appears in McCarthy, 1996:336–338).

Though of an interim nature, Mardikian's model assists conservators and maritime archaeologists to understand concretion processes on iron and steamship wrecks. It is also important in illustrating that within concretions there are regions with differing microenvironments.

In his Stage 1, the trunk, a composite metal structure with numerous galvanic couples, is open to the sea. In Stage 2 the middle section of the trunk became isolated due to the accumulation of concretion and detritus at its once open ends. The heavily concreted diver's knife found in 1994 gives some indication of the speed of this process (Figure 60). In this stage Mardikian identified the effect of sulphate-reducing bacteria, oxygen, carbonate depletion in the water column, a reduction in pH, and an increase in corrosion. In Stage 3 the engine itself is totally concreted, in part protecting, yet at the same time rendering the inner part of the trunk anaerobic, leaving it part-filled with soft, highly acidic corrosion products. His Stage 4 sees the area totally filled, though in some areas the corrosion products are still relatively soft.

While this is but a preliminary analysis, it is an important development, highlighting the need for further research. Moreover, it is only recently that concretions have been shown to be more than just an impediment to shipwreck archaeology. It is now known that they can contain information of considerable significance. For example, microscopic examination of some concretions from *Xantho* has shown that approximately 16 separate bands occurred in their matrix. This indicates that the wreck may have been exposed and buried 16 times since 1872 (MacLeod, North, and Beegle, 1986; MacLeod, 1992a). Information of this nature is of importance, not just to conservators and site managers but also to archaeologists in providing information about postdepositional effects.

8.12 Further experiments in deconcretion

Having proved that the flame deconcretion system was effective in removing very hard or inaccessible concretion, Garcia was requested to begin work on a feasibility study by first removing one bolt on the engine. Heat and lubricant were applied, followed by spanners. The method proved successful and he then removed the bolts on the wrought iron crankshaft. Next, the big-end nuts and assemblies were removed, followed by the cast iron end caps and iron pumps and then the wrought-iron radius links, eccentric straps (XA 396–9) and weigh shaft.

The corrosion under the nuts and bolts was greater than had been expected and it was evident that it would have eventually destroyed these components from within. Consequently, it was decided to remove as many nuts and bolts as possible. When fatigue or other pressures dictated, work stopped and the engine was returned to its bath.

At the end of March 1993, the successes were such that Garcia proposed the engine could (and should) be dismantled into its component cylinders and mainframes for conservation purposes.

At this stage, it is relevant to follow on from an earlier comment about the categories of archaeological information that can result from an examination of a ship's engine. In order to do so, it is reasonable to make an analogy.

8.13 Analogy in engine excavation

In attempting to answer questions raised about any particular machine, an inquirer would turn from the records and verbal statements given by a former owner to an examination of the engine itself. This would first be performed externally by visual observation, then with the machine running. (With the substitution of a working model for the original, to an extent we have done this in the *Xantho* case.) An external examination and compilation of technical data and oral histories would be followed by an examination of the internal state of the engine through the use of cylinder pressure gauges and other sensory aids. Where vital indications are not positive or conclusive, an analysis of the interior of the engine by the removal of the cylinder heads or the covers is possible. If further information is required, or standards are more rigid, such as in the aircraft industry, then the engine could be disassembled and its parts subjected to minute structural analysis, maybe even by metallurgists. In combining all of these approaches, answers could be provided to questions such as: was the engine new or recently reconditioned? Had it been modified in any way? Were those modifications suitable? Was it well maintained? How long would it last? And so on.

These interpretations, when applied to an engine recovered from a past context, have both technical and behavioral dimensions. First, a statement can be made about whether the engine is of a suitable design or whether it was well maintained. Second, conclusions can be made about those who owned and operated it.

8.14 Dismantling the machinery.

When opened the "boiler valve" (XA 339) proved to be a safety valve designed to release steam when the boiler pressures exceeded those considered safe for normal operation. It was a counterweight or deadweight system, one that worked well where boilers are kept in one plane, e.g., on land, on rivers, or in sheltered waters. If fitted to boilers on oceangoing steamships, the valve seat would be subject to varying pressures as the vessel pitched or rolled. As a result it was initially considered anomalous. The type was found to have still been in use at sea up to the end of the century, however (Hutton, 1890). Then the advent of the spring-operated type in Figure 8, one that worked to the same pressure regardless of its plane, ensured that the deadweight system fell out of fashion and was considered unsuitable for use at sea (Sothern, 1923).

The pumps, which were small and joined with few bolts, were also deconcreted and opened up. The starboard pump was completely dry. The two pumps were attached to a *Scotch yoke*, a form of eccentric attached to the crankshaft causing them to be always in operation (i.e., they could not be disconnected). Inside each pump were two nonreturn "mushroom" valves, all originally with a stem around 50 mm long. These were adjusted by a nut on the top of the pump case. In a number of instances the stems had been worn almost totally flat through their continual use. The rudimentary design of the pumps led to the belief that they were not part of the original engine. It was considered that that they were possibly an afterthought fitted by the scrap dealer Robert Stewart. Further, a previously unidentified valve (XA 50) that had been located during the excavation of the stern of the ship was eventually recognised as a spare mushroom valve. It had not been used, and though poorly cast, it provided an interesting indication of the extent of the wear on those taken from the pumps (Figure 70).

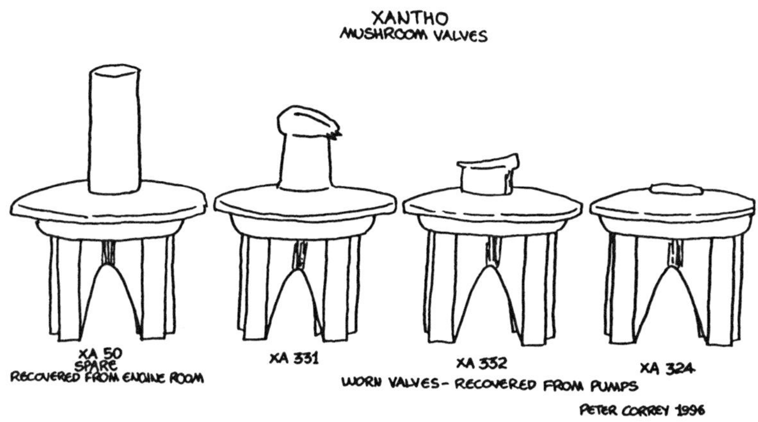

Figure 70. A sketch of three of the mushroom valves recovered from the pumps, showing the extent of wear on those in use in comparison to the spare on the left (Peter Correy).

Having proved a success on the pumps, a small-scale feasibility study was conducted on the engine, with great success. Starting with the fore starboard side, all bolts holding the engine to the bedplate were carefully removed for cataloguing and conservation. This was followed by removing the bolts holding the cylinders to the mainframe, the pump to its bedbolts, and the bolts holding the cylinders together. Each part was then separated under Garcia's direction using hydraulic jacks and levers, tiny rollers or wedges. Slowly the engine was separated into its component parts, just as John Penn had designed. Eventually the cylinders were split into each unit (XA 445 and 446).

Under each of the three webs of the mainframe were a series of rough wooden wedges and iron spacers that were used to align the engine with the crankshaft. The use of spacers is standard practice in engineering, but in this instance they appeared particularly crude. Similar wedges were found under the cylinders. After being recorded and numbered in the usual fashion, these too were removed for conservation. In December 1993, with the main treatment tank near the end of its life (ironically due to corrosion), the engine sections were removed from the tank and placed into smaller containers. Further disassembling of the numerous mass-produced and interchangeable component parts then began.

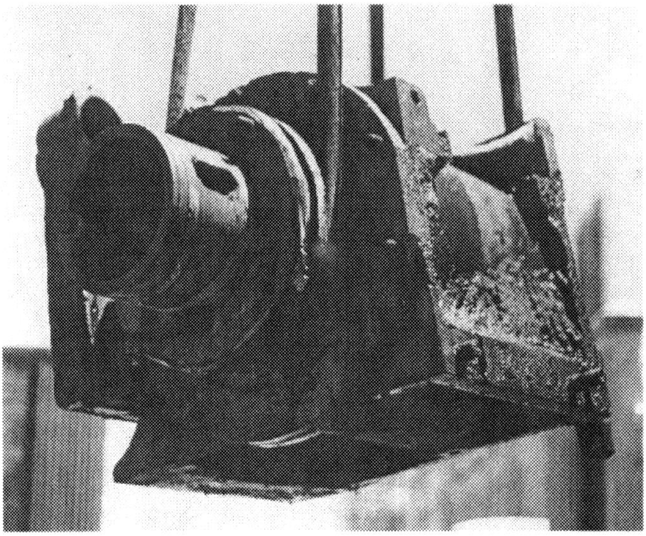

Figure 71. The fore cylinder and valve chest assembly, showing concretion that had earlier proved impossible to remove (R. Gould).

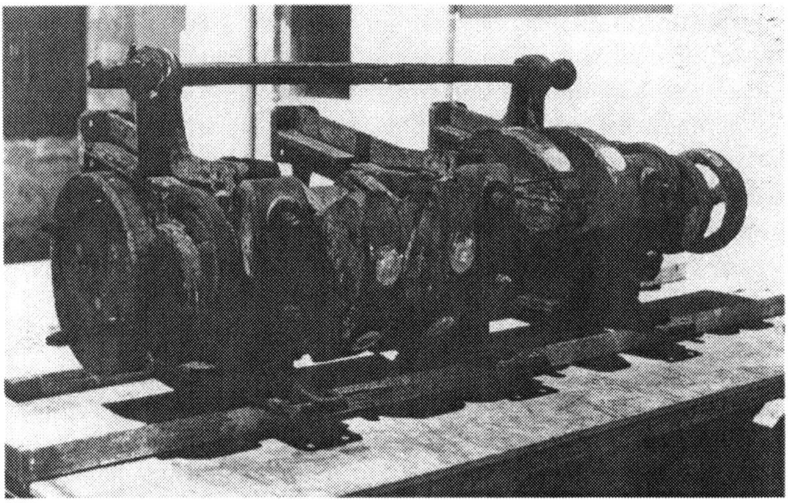

Figure 72. The crankshaft and web assembly. The light-colored circles on the counterweights are lead, which was poured while molten into the recesses to prevent the bolts loosening (Ray Sutcliffe).

8.15 Entering the cylinders

First Garcia commenced removing the glands from the pistons. This required the construction of special purpose mechanical "pullers" and miniature jacks, which were applied, along with heat, to surfaces seized for over a century.

After months of work, cleaning minute channels of concretion and applying heat with tension and lubricants, the steam glands and their packings were removed. In late June the cylinder head was also removed, allowing the packing glands and internal spaces to be recorded and analyzed. The trunk, with its piston attached, was seized inside the piston. Garcia then set about its removal by systematically applying heat, lubricants, and then tension using a specially constructed "puller" and after hundreds of hours the cylinders themselves were entered (Figure 73). Their state reflected the fact that one side of each cylinder was open to atmosphere via the valve chests and the other was steam-and watertight. Sand had been deposited in the open part of the cylinders together with corrosion products and other detritus. The layers were recorded and samples were taken and sent for analysis. The closed sections of each of the cylinders, being steam tight, were in exceptional condition and had no deposits within them.

The interior of the open sections provided very useful information about the events that had occurred in the time after the loss of *Xantho.* Two discrete and opposing lines of corrosion were evident in one cylinder. These indicated that the vessel initially lay heeled to port at an angle of 31° and then settled onto its starboard side 10° from the horizontal. The latter fits the 8–11° range measured at the stern in 1985.

In 1994 the crankshaft was turned for the first time in over a century. Then the end caps, bearings, and finally the crank itself were removed from its frames. These, in turn, were separated and placed in a treatment tank for further electrolysis. The wrought-iron engine bed was also dismantled, revealing that it was in a very eroded and fragile state and barely strong enough to hold the engine.

The deconcreting and total disassembly of the *Xantho* engine allowed it and its component parts to be examined from a technical perspective. As a result, there are many engineering and metallurgical studies forthcoming from an area of inquiry (the excavation and disassembly of an engine after over a century in a highly corrosive environment) that was once considered to be near impossible. Most of these studies are on-going or are awaiting the emergence of the materials from the conservation tanks. Being peripheral to Broadhurst and his crew, most of these technical studies are outside the scope of this particular work, though some of the results may prove relevant.

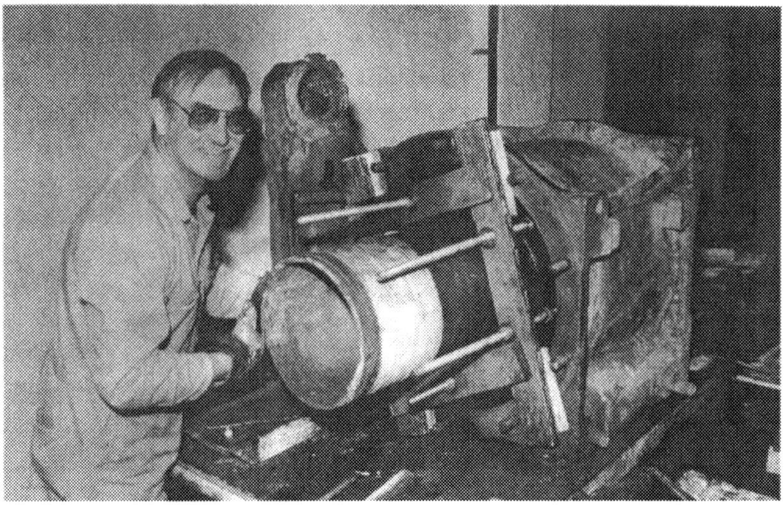

Figure 73. Dick Garcia with the opened cylinders. Visible also is the specially-designed puller (Jon Carpenter).

8.16 The British Standard Whitworth (BSW) thread

One of the observations that resulted from the excavation relates to the threads found on the engine. Even these provide an interesting link to Broadhurst and his family. In 1841 Sir Joseph Whitworth, Broadhurst's future brother-in-law, proposed a standard for the screw thread that was quickly adopted and now bears his name: the British Standard Whitworth

(BSW) thread (Lee, 1900; Gilbert and Galloway, 1978). His career straddled the entire *Xantho*/gunboat period. As indicated earlier, it is possible that Whitworth influenced Broadhurst in his decision to purchase *Xantho*, because he would have been well aware of the pedigree of the gunboat engine type–for we now know that he was a part of its development! Interchangeability of parts required a standardized thread and every thread on the *Xantho* engine has been analyzed—and they are all BSW threads (Garcia and McCarthy, in prep).

The threads on the pumps were also found to be Whitworth standard threads. This was a surprise given the belief that the pumps were not part of the original engine, as suggested by their rudimentary form and by metallurgic studies that showed that the valve chest and the pumps were not cast from the same metal. The possibility that the pumps were an integral part of the engine, but may have been cast and manufactured by a subcontractor in a different foundry, then arose.

8.17 Further evidence of poor maintenance

Finally, in the last stages of the disassembly and deconcretion process during the first months of 1995, we were taken back to Broadhurst and the way he and his men operated the *Xantho*. When the aft cylinder was opened and dismantled, the "little-end" crankpin was found to have been exposed at each extremity by roughly grinding down the casting of the little-end itself. The exposed pin was then apparently heated and hammered over using extreme force at each end; like a rivet.

It appears that the pin had become loose during operation, causing "slap" at the little-end on each stroke, vibrating the engine, and causing obvious concern to the engineers. In order to alleviate the problem, a crude attempt was made to expand the pin *in situ* and to take up the slack, thereby reducing vibration and wear. The engineers had shaved down the metal sleeve through which the pin was fastened to the little-end, thereby exposing around 1–2 centimeters of the pin at each end. The exposed pin was then heated until malleable and "peened over" using heavy hammers.

In dismantling the piston and removing the rings, it was found that one of the nuts holding a locking ring in place on the face of the piston had worked loose and had fallen into the drain plug at the bottom of the cylinder. The locking ring was working loose and had it done so, the rings and piston would have separated and component parts could have been released into the cylinder midstroke, with dire consequences.

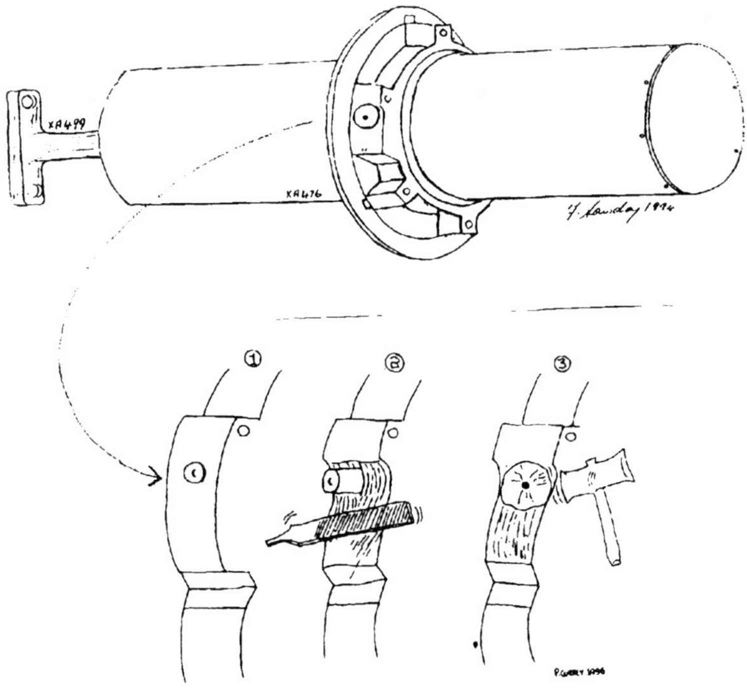

Figure 74. Preliminary sketches of the trunk assembly, showing the position of the little-end gudgeon pin and the three stages in the "peening over" of the pin using hammers and heat. (Fairlie Sawday and Peter Correy). (Not to scale).

Thus the *Xantho* engine was patently succumbing to the combined effects of high speed, high working pressures, and poor maintenance. These were precisely the kinds of problems that caused the British Admiralty to station a special purpose, properly equipped floating machinery workshop to service the gunboat engines when in naval service. Technology essential for one context can easily fail in another, due to the absence of ancillary support facilities, expert staff or to the presence of factors outside its normal or planned operating parameters. Broadhurst and his crew could not have been unaware of the fact that their engine was in dire need of expert repair and refurbishment. Their problem was, where and how?

Thus, though there was some logic in operating a steamer fitted with a simple, compact, mass-produced engine with an apparent abundance of spare parts, there were still enough flaws in its final application to cast doubt on Broadhurst's powers of judgment. The hull was worn out, the pumps were on the verge of breakdown, and the boiler was of a high enough pressure to cause heavy precipitation of its saltwater feed. That the engine was falling apart, poorly maintained and running in reverse in order to drive the vessel forward, as a result of the fitting of an incorrect propeller–only compounded the problems.

8.18 A setback in the final stages

Following the full disassembly of the engine by April 1995, the various components were collected and returned to new set of treatment tanks with new anodes and other accoutrements. The quantity of salts leaving the newly-exposed surfaces led Ian MacLeod to observe that "at this stage after nine years of the project it felt as if we were just beginning again" (*Xantho Engine Deconcreting Book* April, 1995). To start again, no matter how frustrating, was an essential step, for it had become evident throughout the disassembly process that if the engine had not been reduced to component parts it would inevitably have been destroyed by corrosion from within. The treatment proceeded smoothly for another two years and all was in readiness to receive the engine parts for reconstruction in an exhibition gallery.

Then, in February 1997, there was extensive exfoliation of many of the cast iron parts of the engine over the course of one weekend. After the team recovered from the setback, a review of the situation commenced. Neil North was commissioned to conduct an external analysis and in an effort to uncover the cause, all the steps taken from the time the engine was found were revisited. In respect of the recovery and transport of the engine, he considered it "difficult to imagine any other means by which this type of artefact could be managed" (North,1997:8). In detailing the problems and their causes and in making numerous useful recommendations and observations, North concluded thus:

> The majority of the problems experienced in the *Xantho* engine treatment have arisen from the long time required to effectively deconcrete and dismantle the engine, during which the metal was exposed to the atmosphere and corrosion attack commenced. These have to some extent been compounded by other factors, but even if these had not existed it is likely that the end result would be the same as the current situation. When projects of this type are undertaken in the future, considerable attention will need to be paid to development of alternative deconcreting methods which do not result in extensive atmospheric exposure. This is not simple. The methods currently used, which were developed successfully for relatively small and simple artefacts, have proved to be inadequate for large complicated artefacts. . . . Other methods of deconcretion, such as use of inert atmospheres and deconcretion while immersed, may be practical for small artefacts but have major problems, and operator safety considerations, when applied to large artefacts (North, 1997:14).

An examination of the flame deconcretion method was also undertaken, but no correlation between the application of the method and the items suffering exfoliation was apparent. It was evident that the cast iron counterweights on the crankshaft suffered some of the most severe damage, for example, yet these had not been subjected to the flame deconcreting method (North, 1997).

Ian MacLeod's analyses indicated that an undetected electrical separation was present between the corroded surface zones and the residual metal under the graphitized layers. This had developed during the deconcreting sessions in the 18 months to 3 years after the engine was recovered. The loss of direct electrical contact between the external graphitized surface and the underlying solid metal had apparently been the result of the application of a freshwater spray to the engine without a corrosion inhibitor and/or water dispersal agent to the deconcreted surfaces. MacLeod believed that measuring the corrosion potentials on the external graphitized layers of the deconcreted surfaces, rather than internally on the core metal itself (especially when they were not in direct electrical contact), was the major cause of the problem. This gave false indications of the underlying situation and resulted in the application of treatment voltages of a magnitude that resulted in the evolution of excessive amounts of hydrogen gas at the interface of the residual metal and the graphitized layer over the weekend (MacLeod, 1999).

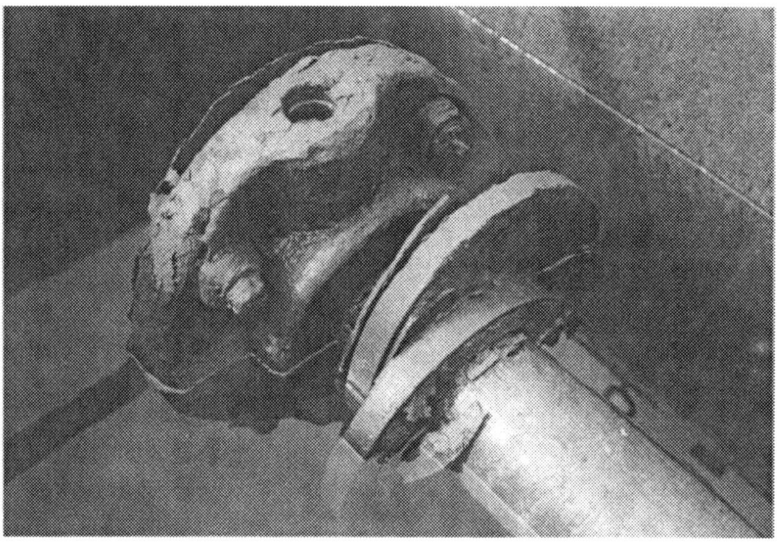

Figure 75. Exfoliation at the aft end of the crankshaft (Jon Carpenter)

The analyses mentioned above and this particular work are presented partly for the benefit of others interested in building from the experiment in conservation and archaeological method that the deconcretion and disassembly of the *Xantho* engine represents. By presenting both our successes and misfortunes in this manner (as a predetermined element of the *Xantho* research strategy) it is hoped that a suitable foundation for similar studies has been prepared.

Chapter 9

Conclusion

Description, analysis, and explanation at Xantho

9. IRON WRECKS AS ARCHAEOLOGICAL SITES

Research conducted underwater at the site of SS *Xantho* has further illustrated that iron and steamship wrecks are a finite and legitimate archaeological resource.

An experimental and innovative steam engine, the product of a strategic need on the part of the warring British government, became redundant, was taken outside its normal operating parameters, and was unsuccessfully reused in a remote colonial context.

Lost in warm, highly oxygenated waters, refound, and then considered an endangered object (partly as a result of being found), the heavily concreted engine was recovered from a saline environment after a century on the seabed. Then it was successfully deconcreted and totally disassembled for technical and archaeological purposes. A great deal of 'lost' or unavailable information about the manner in which the *Xantho* was engineered and about each specific piece of machinery, especially the engine, has since been obtained.

9.1 Description and analysis

While the use of on-site conservators in maritime archaeology was not new when the *Xantho* study commenced, the involvement of conservation specialists in *ab initio* underwater analyses was a logical progression from the earlier practices. The program combined archaeologists and conservators on the seabed from its beginnings in 1983 and then treated the two as a single team for archaeological, as well as conservation, purposes in all subsequent stages—both underwater and in the laboratory. As a result, it could be argued that the project provided a new direction in the study of iron wrecks in the 1980s (cf., McCarthy, 1986d, 1989). At the time it was claimed that

> The wreck site of the iron steam-ship *Xantho* has provided a model for how an underwater archaeological site can be managed. Predisturbance surveys of the marine biology and electrochemical and physical environment of the site established reference criteria for monitoring changes in the site conditions (MacLeod, North, and Beegle, 1986:113).

In utilizing the services of these specialists, the *Xantho* project has been able to proceed in a scientific fashion in examining postdepositional processes. The practice has become more prevalent in recent years on wrecks and lately at port-related structures such as jetties (e.g., Carpenter and Richards, 1994; Garratt *et al.*, 1995; Gould, 1991; Guthrie *et al.*, 1994; Kenderdine and Jeffery, 1992; Lenihan and Murphy, 1989; MacLeod, 1992b; 1998a & b; McCarthy, 1998c; and Murphy, 1987; 1998).

Shipwreck corrosion scientists have since recognized that the principle of applying anodic protection devices, or impressed current, to a historic wreck has considerable merit. First applied to *Xantho* in 1983, initially a failure due to their unexpected burial in sediments, the anodes were re-applied to the engine in 1984 and to the stern in 1985. They were monitored over the ensuing decade, proving (when not buried) a success in commencing the *in situ* conservation of the stern. The process has since been repeated elsewhere on anchors and cannon and lately at the 167-foot-long (50.9 m) iron barque *Santiago* (1845–1955) in South Australian waters (Kentish, 1995). Recently, they were connected to the submarine *Resurgam* and their effects are being analyzed by David Gregory of the Danish National Museum's Marine Archaeological Research Centre.

Following on from the involvement of Maria Pitrun in the analysis of the material raised from the *Xantho* wreck, archaeologists have begun to scientifically address (albeit inconclusively) the question first posed by Larry Murphy (1983)—whether it is possible to show that predepositional corrosion or stress had led to the loss of a particular ship (e.g., MacLeod and Pitrun, 1986; MacLeod, 1992a).

Maritime archaeologists and corrosion scientists have also come to appreciate the vast amounts of archaeologically relevant information that even mere rust stains contain. Indeed, as Turgoose noted with respect to corrosion products and the information they contain about site formation processes, removal of it when necessary, should be performed with a clear awareness that "potential information about the artifact and its burial environment are being lost"(1989:30–31).

Finally, with respect to description and analysis, in establishing what were commonalities in iron and steel ship disintegration and in examining these against the case of the *Xantho,* anomalous features were identified. In this manner, the *Xantho* program has assisted in the development of general (though, as yet, scientifically untested) models for iron and steam shipwreck disintegration following on from Muckelroy and Riley. By controlling for these postdepositional effects informed comment was made on both the abandonment processes and such fundamental questions as whether the loss of the vessel was deliberate or accidental.

When the *Xantho* results are combined with those from contemporary studies at the USS *Monitor* and the USS *Arizona* mentioned earlier, it is evident that iron, steel and steamship wrecks are rapidly deteriorating. When these results are read in conjunction with subsequent corrosion studies it becomes apparent that the recording and in some cases the protection of iron, steel, and steamship wrecks should receive a high priority. On going studies, apart from those mentioned previously, are research at the SS *City of Launceston* (1863–1865) in Port Phillip Bay, Victoria (Strachan, 1999), and an entire suite of wrecks in the waters of South Australia, including the turret steamer *Clan Ranald* (1900–1909) (Arnott, 1999; MacLeod, 1998a). In some of these cases deterioration has been quite noticeable or was visibly accelerated by human intervention. In comparing data in the years from 1979 to 1990, it was understood that the rate of deterioration at *Monitor* was rapidly increasing, for example (Arnold *et al.*, 1992).

In noting that "iron will not last forever," one influential practitioner, Nic Flemming, echoed one of the major reasons why the study of iron wrecks, even modern steel ones, is presently warranted. He estimated that the wreck of the *Titanic* will "probably crumble within a further 75-100 years" (Flemming, 1988:198–200). Here reference is made to the situation above the seabed and it needs also to be acknowledged that there is still a great deal of research to be carried out on what occurs beneath the sea floor. Microbiologists (Guthrie *et al.*, 1994) have conducted a survey of postdepositional and post-excavation microbiological effects on the wooden-hulled sailing frigate HMS *Pandora* (1779–1791). Though this initiative was preceded by an analysis of biologically induced corrosion at the late eighteenth-century Yorktown Site in Virginia (Rodgers, 1989), it too represents a very significant development. It is doubly important, for apparently there still exists a lack of "valid experimental procedures allowing the independent measurement of both the electrochemical and biological components of corrosion systems in such a manner that the nature of their interdependence is made manifest" (Sequeira and Tiller, 1988:17).

In recent years there has been an increase in the number of practitioners and a resultant series of important advances in on-site shipwreck corrosion studies. David Gregory's analysis of the use of anodes at the *Resurgam* and his critique of corrosion measurement methods and the use of anodes on cannon at the Cromwellian Duarte Point wreck, are but two very important examples (1999). Another important step has been the use of dataloggers on wooden wrecks in Britain (Merrett-Jones and Pedley, 1998) and suggestions in the electronic media on the possibility of their use on iron and steamship wrecks. Notes on the application of specialist expertise and equipment have become a regular feature of discussion lists such as the well-known "Subarch" and now "Seasite" (http://www.mailbase.ac.uk/lists/sea-site). Other initiatives, such as a "process-oriented approach" attempting to model for different rates and degrees of physical, biological and chemical processes at the remains of HMS *Pandora* (Ward, *et al.*, 1999) may also prove of significance to iron and steamship archaeology.

Finally, in respect of the ethics of removing complex objects such as marine engines from the seabed, the excavation, disassembly and conservation of the *Xantho* engine were new procedures, each requiring considerable experimentation. The *Xantho* engine was the second piece of marine machinery to be removed for study, conservation, and exhibition purposes, for example. It was preceded by the Propeller *Indiana* engine, which was partly disassembled soon after it was recovered from the relatively benign waters of the Great Lakes. Most of the *Indiana* machinery now lies in store, having required little conservation treatment (P. Johnston, personal communication, September, 1994) and the turning screw forms the centrepiece of an impressive exhibit at the Smithsonian in Washington, D.C.

The third piece of machinery removed from its maritime archaeological context was from the steamer *Arabia* (1853–1856). Parts of its propulsion system were recovered in 1988 and are now on exhibit in Kansas City (Hawley, 1989). This was followed by the crosshead engine of the Steamboat (PS) *Columbus* (1828–1850). This machine was totally disassembled after being recovered from the saline waters of Chesapeake Bay in May 1993 (Holly, 1996). Though some parts were in need of urgent remedial attention when visited by the author in late 1999, this engine lies in the best possible climatically controlled conditions at the extremely well-equipped Maryland Archaeological Conservation Laboratory. In being totally disassembled, it also provides an opportunity to examine the manner in which it was operated and it has the potential to be a most remarkable exhibit indeed.

The results in these four cases are mixed. Despite this, indications are that if the problems encountered with the *Xantho* engine in the penultimate stage of its conservation can be avoided, then to consider the recovery of a historic marine engine, especially one that is endangered, is a valid exercise. Clearly there must be sufficient reason to do so, however, and the appropriate facilities, staff, finance, long-term commitment, and expertise need be obtained beforehand.

A step forward in iron and steamship archaeology post-*Xantho* is the detailed preparation for the raising of the Confederate wrought-iron submarine H.L. *Hunley* and its crew. This has included the prior obtaining of finance, social and political backing, expert staff, a large conservation facility and an existing and well-regarded exhibition venue in Charleston, South Carolina. The convening of an international forum designed to air the many conservation, ethical, and archaeological issues, with a view to their satisfactory resolution well before the excavation and lift occurred, was another step toward the establishment of best practice in iron and steamship archaeology. Papers presented at the November 1999 seminar are being compiled (*Hunley* Team, in prep.).

This situation contrasts with the recovery of the German submarine *U 534* from Danish waters in 1993 (McCarthy, 1998b). Though archaeologists and other specialist were apparently employed in the excavation of the wreck, there appeared to be no prior long-term national or institutional commitment to its conservation. It was then transferred to Birkenhead in England and is now presented in a historic location alongside other heritage vessels. When visited in November 1999 it proved to be an evocative and most worthwhile experience.

It was also evident that, though there exists a dedicated and energetic staff at Birkenhead, they are in urgent need of a massive influx of capital, equipment, expertise and materials in order to prevent *U 534* deteriorating further. Excavation was still incomplete, for example, and the conservation of all, except the external hull, has been rudimentary.

Without a great deal of support this important relic may be put at further risk. Despite these concerns, the excavation and exhibition of *U 534* is proving to be both an important development and a cautionary lesson in maritime archaeology, museology, and conservation science. These lessons are directly relevant to the present debate about raising, excavating, conserving and exhibiting both *Resurgam* in Britain and the historic Australian submarine *AEII* (1914-1915) in Turkish waters (Smith, 1998; Spencer, 1999).

9.2 Analysis and explanation

In answer to the question of whether an already well-documented iron or steamship wreck is capable of adding to the understanding of the social context and behavior of its owners and operators, it is now clear that the answer can be in the affirmative.

The potential relevance of iron and steamship studies to anthropology has been established in the *Xantho* project, where it was attempted to define physical parameters before analyzing the behaviour of the owner Charles Broadhurst. As distilled from the definitions of Muckelroy (1978, 1980), Murphy (1983) and Renfrew and Bahn (1991), it is evident that research at iron and steamship wrecks is capable of elucidating otherwise unattainable aspects of behavior, thereby shedding new light on past human life.

In 1983, the year the *Xantho* project commenced, Daniel Lenihan noted that the questions put by marine architects and maritime historians "are different from, but every bit as valid" as those asked by anthropologists (1983:43). Though acknowledging the importance of both approaches, Patty Jo Watson summarized the underlying tension between historical particularism and anthropologically based approaches to maritime archaeology when she noted that, while each will stress one approach ahead of the other, both points of view are "essential and both are present in everyone's work" (1983:310). The treatment of 26 diverse shipwrecks in the Isle Royale National Park in Lake Superior under Lenihan's direction reflects this point of view, and in his words, it "represents an attempt to accommodate the best from history and social science in a cultural resources management framework" (Lenihan *et al.*, 1994).

The fact that archaeological theory is evolving quite rapidly became evident over the term of the now 17-year-old *Xantho* program. Where possible the author listened and took objective note of each new development and responded accordingly. From a processual perspective, for example, Peter Veth considered that the *Xantho* could be seen as a component of entrepreneurial expansion on a major frontier underwritten, though sometimes impeded, by bureaucracy and the legislature. The vessel spearheaded an ambitiously broad range of economic enterprises, including pearling, whaling, fishing, and the carriage of passengers and general cargo over a vast geographical area ranging from Fremantle to Batavia (Jakarta). Such diverse endeavors arguably required a non-specialised craft capable of optimising returns on any potential commodity. Given the frontier setting, the craft would have had to be simple and robust, of low maintenance, with available spare parts. It also needed to be capable of operating with a paucity of resources and a minimum of infrastructure. One hypothesis that could be considered is the question whether such optimizing strategies as those exhibited in the *Xantho* case are a feature of entrepreneurial maritime groups in the initial colonisation of coastal frontiers (see discussion on page 190).

On the other hand, a cognitive processual critique might view the numerous anomalies identified at the wreck of the *Xantho* as symbolic of, or representing, the unconventional and idiosyncratic approach of Charles Broadhurst, the unpredictable individual. Further, a post-processualist might seize upon Broadhurst's curious behavior with the *Xantho* as a real example, supported by the historical record, of the impossibility of applying, or even trying to apply, general rules or semiquantitative analyses to the human situation, especially where enough funds exist to fuel capricious whims.

There are a number of reasonable possibilities and all need be explored before settling on that which is best supported by the available evidence. In the purchase of the aging and ill-engineered vessel and in its mode of operation, for example, Broadhurst may have been exhibiting commonly acknowledged failings of those born into privilege and power—access to money, power, and influence, with a surfeit of ideas, but often without the practical skills and business acumen required to bring them all together in a productive manner. The gulf between Broadhurst and his contemporaries, such as the very successful colonial entrepreneurs Walter Padbury, an orphan child (Nairn, 1984) and Charles Harper (Mercer, 1958), a local boy who grew up with Aboriginal people, is stark. Broadhurst was markedly different in both his approach and his methods in the same environment. To further illustrate this point we may look at the technology found at the *Xantho*. There, a curious marriage was found of a hull designed for sheltered waters with a mass-produced, noncondensing, seawater fed, high-compression, high-revolution, energy-expensive steam engine running in reverse (due to an incorrect propeller). The engine was also showing evidence of very shoddy maintenance. This makes sense if we see the purchase of the vessel as the product of poor engineering advice and possibly Broadhurst's own naiveté, eccentricity, and self-belief. As the archival sources show, when he had good advice and good managers, he succeeded—when he did not he failed. He also attempted to apply otherwise useful technology out of its context.

On the other hand, Broadhurst was an undoubted visionary, continuously attempting to apply new ideas and technology in frontier environments, a man with undeniable drive and a remarkable capacity to interest others and to raise some of the funds necessary to pursue his dreams. His object in acquiring a multipurpose carrier capable of operating in the face of all kinds of material shortages, with a very simple engine having interchangeable parts, capable of operating without freshwater and away from engineering facilities, can also be seen a result of an identifiable and logical process.

Though he never appears to have admitted it to officials or to the public, Broadhurst may have intended to operate *Xantho* as an "auxiliary steamer" (or "steam schooner") i.e., a ship where the engine was secondary to the sails. One solution to his problems in operating the machinery in such a difficult environment could have been to use steam only when absolutely necessary, or on public occasions when attempting to show *Xantho* off to advantage. If this was so, then the condition of the engine was not crucial to Broadhurst's plans.

Further, the idea of using a steamer in the northwest is traceable to his earlier involvement in the Camden Harbour Pastoral Association and the Denison Plains Pastoral Company. Both of these endeavors flagged the benefits of an independently-operated small steamer that would link the far northwest coast of Australia with the outside world and thereby provide a new gateway to this continent. One major failing, in the context of Broadhurst's visionary zeal, can be attributed to his lack of attention to the minutiae required for successful steamship operation in such a remote setting. These details might include the inappropriateness of the scantlings and operating parameters of the vessel, the availability of repair facilities and coal supplies and, ironically, the maintenance of his insurance cover.

Larry Murphy's "one-more-voyage" hypothesis is relevant to this discussion (Murphy, 1983:75). Broadhurst could not have been unaware that his ship was ailing. The engine would have been vibrating badly, the hull was disintegrating before his eyes; yet he carried on, even to the point of knowingly overloading the ship on its last voyage. He either saw that its loss was inevitable and was going to squeeze the last ounce of use from it before it sank (risking crew, cargo, and himself), or he had a misplaced confidence in engineering.

Arguably one basis for the one more voyage syndrome is the availability of insurance, in that the taking of risk can be directly related to access to insurance in its various forms. What clouds Murphy's analysis of shipowner behavior and its applicability to the *Xantho* case may be the failure to reinsure it as Broadhurst had intended. In contrast, Richard Gould recently indicated that this phenomenon i.e., operating uninsured vessels to the point of failure, is also a manifestation of Murphy's one more voyage syndrome. He indicated that operating uninsured vessels is just another form of high-risk behavior on the part of unscrupulous shipowners– in other words, Broadhurst may have been hiding the fact that he deliberately did not insure *Xantho* and was prepared even to see it sink in wringing every possible ounce of use from it. (R. Gould, personal communication, April 3, 1996). That Broadhurst was prepared to "save the cargo rather than the ship" as it sank beneath him at Port Gregory (see page 56) provides an indication of this. Research recently conducted at Fort Jefferson in the Dry Tortugas (Gould, 1995; Souza, 1998) indicates that shipowners generally operated vessels to that port even after they had been condemned and the insurance cover withdrawn, for example.

Broadhurst's role as an innovator ahead of his time, or the notion that inventions sometimes have to "wait their time", may also be examined at this point. Daniel Lenihan, for example, proposed that "certainly shipwrecks over time offer an excellent database for getting at this question" (1983: 56). This concept is relevant both to an analysis of Charles Broadhurst's personal propensity for innovative methods and new ideas and to the technology evident at the wreck of the *Xantho*. Broadhurst consistently applied technology well ahead of its time (e.g., the application of diving apparatus and the use of steam power in the pearling industry).

It is evident to the author that Broadhurst, the Victorian-era gentleman born into wealth and command, had vision but lacked the practical focus and common sense necessary to translate ideas into a viable enterprise. He clearly exhibited the "misplaced confidence in engineering" that Gould saw as common among men of the Victorian era (1990:55). He also exhibited what Gould (1990:54) has identified in other circumstances as social and cultural factors that help account for particular high-risk behaviour.

As a result of his background and his fall from grace and social position following the dramatic failure of his foray into the Denison Plains Pastoral Company, Broadhurst's personal failings were dramatically bared. He appears from then on to have become singularly obsessed by a search for wealth, security, and position. His precipitate and well-publicized social demise may be seen as the major driving force behind his subsequent high-risk behavior. This manifested itself in his acknowledged propensity to "go out of the ordinary grooves in search of wealth" (Kimberly, 1897:97), to travel vast distances, to risk himself and his family, and most pertinently, tc experiment with untried technology and ideas in a frontier environment.

One could also argue that this propensity represents a pattern where a wealthy and socially well-placed individual with very high familial, social, and financial expectations is driven to perform extraordinary feats in order to resurrect a destroyed career or social position. The only alternative for him was unaccustomed mediocrity, embodied in the government position he sought when at his lowest ebb (see page 122).

Further, the effect of his misfortunes, failures, and endless controversies on Eliza was considerable and was acknowledged by him on a number of occasions (see McCarthy, 1990). Whatever his motivations, Broadhurst experienced phoenix like resurrections far too often for them to be a coincidence.

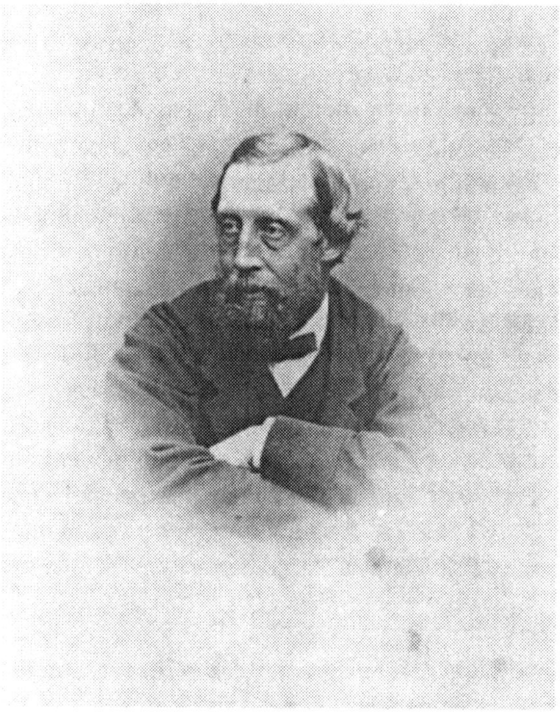

Figure 76. Charles Broadhurst (The Broadhurst family).

Finally, in searching for behavioral generalizations emanating from a broad-based study of iron and steamship wrecks that might follow on from the SS *Xantho* case, the following propositions were made with the assistance of Peter Veth (Veth and McCarthy, 1999).

- Vessels used by individual entrepreneurs in frontier contexts tend to be non specialized and simple.

- In frontier settings engines and general mechanical fittings will be selected for low maintenance and ease of interchangeability of parts, rather than for efficiency.
- The "robustness" of such vessels is a reflection of deliberate redundancy in that numerous aspects of the *Xantho*'s engine (as one example) were aimed toward replication and interchangeability. The vessel (and therefore the system) is less likely to fail should a single component fail.
- Craft owned by individual entrepreneurs as opposed to corporations tend to have higher rates of failure.
- Frontier craft may be designed in such a way that they can be used for a wide range of carrying functions and specifically for a variety of functions that might not be envisaged at the time of initial use of the craft. For example, the layout of the interior of the hull *Xantho* allowed for radical redesign of the internal configuration of cargo space. As has been noted, the location of the compact, high-pressure engine on the *Xantho* allowed for maximum use of cargo space.
- When vessels are owned by individual entrepreneurs there may be a greater potential for innovation and, therefore, discard of inefficient features following failure. The opposite case would be the continuing presence of obsolete features such as ramming devices on ironclads well after they were demonstrated to be ineffectual, because corporate groups (such as the admiralty) were locked into what has been described elsewhere as "trend innovation" (cf. Gould, 1990:170 *et seq.*).

In examining these generalizations, it is pertinent to note that there are remarkable similarities between *Xantho* and the wreck of the iron-hulled SS *Sunbeam* (1861–1892), illustrated below and in Figure 34. It was built of iron in 1861 on the Thames and was a 92.1-foot-long (28 m), 72-ton, clincher-built, one-deck, three-masted pleasure schooner. Its original engine was a vertical single-cylinder annular engine built by John Penn's chief competitor, Maudslay, Son and Field. After 25 years on the coast of Britain under numerous owners, *Sunbeam* was re-engined with a two-cylinder 18 hp compound engine and was placed on the market, most likely (given the size of the engine) as a "steam schooner" or "auxiliary steamer". The newly refitted *Sunbeam* was purchased by the well-known pearler and entrepreneur Edwin Streeter for use in the Australian pearling industry (Henderson and Sledge, 1984; Streeter, 1886; Stanbury, 1994).

Like *Xantho* it had only one successful season, in this case 1891. When lost in March 1892, it was operating with a crew of four Europeans and 35 "Asiatics" (as they were called), utilizing four boats with diving apparatus. While at anchor in shallow protected waters near Osborne Islands just north of Camden Sound in the far north of Western Australia (Figure 13), a leak was discovered and the vessel was run ashore to be patched up. It was refloated after the repairs were effected, but 12 hours later the first mate reported that it was again taking water. The master, on going down into the engine room, found a 14 centimeter gash in the hull caused by corrosion. The torrent was quickly plugged and they tried to run the vessel back ashore, but they struck a sandbank en route and the ship became firmly embedded in it. The *Sunbeam* slowly settled down to the seabed, coming to rest in the sediments at or around its waterline (Sledge, 1978). There it remains today, looking (but not necessarily being) completely intact. Corrosion potential measurements have yet to be taken and biologists are as yet to comment on the nature of the colonizing marine life.

While indications are that corrosion and old age were the cause of the loss of both *Xantho* and *Sunbeam*, it is also interesting to note that, like *Xantho,* the wreck also had a known effect on indigenous people. *Sunbeam* has actually become part of the Gambera people's (Horton, 1994) legend or *dreaming*. They understand that the vessel sank as a result of the intervention of a supernatural force or Spirit that was called into action by aggrieved Aboriginal men in retribution for the sailors transgressing against their social customs in respect of the Aboriginal women (I. Crawford, personal communication, November, 1996). Being thereby unwelcome in such a remote frontier situation, the crew were forced to make the journey to Broome in a small boat.

In respect of its role in the development of the northwest pearling industry, the following was said of *Sunbeam*:

> The *Sunbeam* added an important element to the industry-steam power. . . . If the *Sunbeam* had not been wrecked, it might well have set the pace and led to earlier mechanization of the industry–at least for the mother-ships operating in the most remote high-tide areas (Henderson and Sledge, 1984:28–32).

While there appears to be no historical connection between Broadhurst and Streeter, *Sunbeam* or *Xantho*, we now know that the words contained in the quotation above actually apply to startlingly similar events that took place two decades earlier with *Xantho* and Broadhurst. They also indirectly illustrate a commonality of behavior on the part of two wealthy frontier steamship owners and two acknowledged pioneers in the pearling industry.

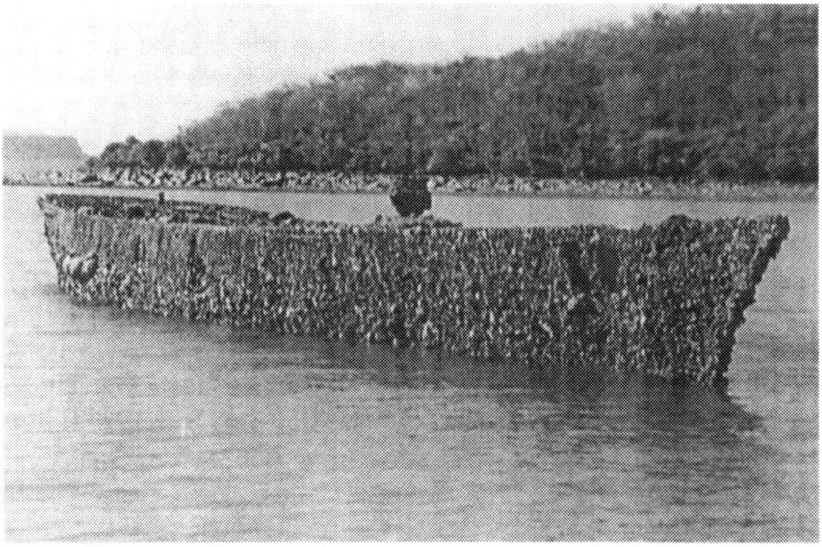

Figure 77. SS *Sunbeam*, showing its hull and the proximity to the Gambera people's land (Western Australian Maritime Museum).

What is interesting in these remarkably similar events is that both men were noted entrepreneurs and that both had training and experience elsewhere; Streeter as a jeweller, Broadhurst as a pastoralist. Both were Victorian-era gentlemen, with social standing and access to finance. One succeeded and was lauded in numerous contemporary and subsequent accounts, including his own work (Streeter, 1886). The other was shrouded in controversy and failure, and until the present study was commenced, was discounted by all, including his family. Like Streeter,

Broadhurst was a gentleman accustomed to position and wealth, an entrepreneur with access to some funds; but unlike Streeter it appears that he never had quite enough to properly cater for all his needs.

Broadhurst also had a propensity for grand schemes and untested technology and precipitously embarked on speculative ventures without the capital, experience, or common sense required to make them a success. He consistently failed, primarily due to his lack of experience, poor advice, and consistent lack of attention to detail. A key is found in his managers. Where they were good, he succeeded; where they were bad, or gave poor advice, he failed. This may be related to his traditionally poor labor relations in that by alienating his workforce Broadhurst may have set the scene for his ultimate demise in most of his business ventures. Nevertheless, he was remarkably resilient. Each time he failed he rebounded through flair, vision, and hard work, only to fail again. When he did achieve some measure of success, he immediately looked elsewhere for other opportunities.

In relation to the extraordinary breadth of Broadhurst's activities; in maritime archaeology, what is found at sea will generally be linked to what is found on land (e.g., McCarthy, 1998d). Toward that end, a survey of some of Broadhurst's Abrolhos Islands guano sites has been undertaken (Stanbury, 1993). The historical archaeological study of pearling bases at Shark Bay with emphasis on Wilyah Miah, where Broadhurst was once based, is also under way (McGann, in prep.). A preliminary examination of one of Charles and Eliza Broadhurst's main pastoral bases has also been made and his pearling base at Banningarra, east of Nickol Bay, was recently examined as part of a study into port-related structures in Western Australia (Cumming *et al.* 1995). The study of Broadhurst's almost-forgotten role in the fish-canning industry at Mandurah has also begun (McCarthy, 1990).

9.3 The *Xantho*/Broadhurst exhibition

As indicated earlier, the SS *Xantho* program is museum-based, having a public access and an education brief. As a result, the site is part of a recognized wreck access facility and interpretive material has been widely disseminated throughout Western Australia. Exhibition of the materials raised and an analysis of the career of *Xantho* and the life and times of Charles and Eliza and Broadhurst is on going. Comment is made in the galleries on the activities of their son Florance Constantine Broadhurst, the guano merchant, and of their daughter Catherine Elime Broadhurst, a suffragette, who chained herself to the London railings. She was then imprisoned for views whose seed can possibly be traced to her mother's trials and tribulations as a Victorian-era woman. For whom cultural difficulties were exacerbated as her husband ranged far and wide in search of wealth and social redemption. That Catherine required force-feeding while on a hunger strike is some indication of the strength of her feelings on women's issues.

There is also interest in developing an exhibition on the subject in both Fremantle and Geraldton and there has been a proposal to travel elements of it around Australia.

The engine is to be the center-piece of the museum's exhibition, despite the problems encountered in March 1987. Ian MacLeod recently finalized successful experiments with an inert synthetic thermoplastic resin that has allowed the exfoliated fragments of the *Xantho* engine to be reattached or reconstituted. Though acknowledged as not ideal, the successes have been such that in April 1999 the team began rebuilding the engine in the exhibition gallery, allowing the public to view the process over the three to five years the rebuilding is expected to occur.

By the time of writing, the brasswork was on exhibition along with the working model and film of the excavations both at the wreck and in the laboratory. The wrought iron engine bed has been reassembled and is now mounted on a frame in the gallery. Conservators are working in the gallery applying finishing touches to cast iron pipes, hardened steel piston rings and other objects. Volunteer historians, assistants, and researchers are also based there from time to time. By the time of writing, the cylinders, valve chests, webs, and cranks still await the completion of the conservation process, however.

Initial public reaction to the exhibit and to scrap-metal merchant Robert Stewart's engine bed—the first item to be rebuilt and placed on exhibition in the gallery—is surprise at its flimsy and rudimentary nature and disbelief that Broadhurst could have been so easily misled in purchasing the ship. This discovery and the exhibition's focus on Charles, Eliza, and their family takes the viewers back to what eventually became the primary focus of the *Xantho* project–the people involved, their behavior, and the motivations for it.

Appendices

10. APPENDIX 1: HORSEPOWER

Note: Over the years and across national boundaries, horsepower (horse power or Horse Power) is seen abbreviated as HP, H.P., and hp in texts, with terms such as Indicated Horsepower (as but one of its derivatives) appearing variously as i.h.p., IHP, I. H. P. and ihp.

Style manuals and editorial guides sometimes present a variety of abbreviations even within their own text. *The Oxford Dictionary for Writers and Editors* (1989) abbreviates horsepower as hp, yet it shows indicated horsepower as i.h.p., for example. Within this work (unless using a direct quote) this author has elected to use hp, ihp, and nhp.

There are also many contemporary and modern definitions of horsepower and its derivatives. Their perceived suitability can reflect the progress of marine engineering at the time of writing and the understanding (and needs) of the reader. Having cast around for a suitable "simple" set of definitions, the following compilation from a variety of sources was developed by the author.

Around the time steam engines were coming into vogue and began replacing muscle to perform a myriad of tasks, the recognized scientific unit of work was the "foot pound", or the force required to move 1 pound-weight (lb.) over a distance of 1 foot. The power (or rate of performing work) of a machine or contrivance could be measured in foot pounds of work performed in one minute. When applied to heavy tasks this became a very large figure, especially with the new breed of heavy steam engines coming into vogue in the nineteenth-century. It was also a figure that the uninitiated had difficulty with. Because dray horses once performed many of the tasks of the large steam engines, engine makers came to express the power of the engine in the more simply understood terms of the equivalent number of dray horses required to do the same work or horsepower (hp).

According to nineteenth-century engineer Professor Andrew Jamieson, the inventor James Watt (1736–1819) established the unit of "horse-power" in a series of experiments conducted at a London brewery, finding a heavy dray horse capable of lifting 22,000 pounds a distance of 1 foot in one minute (1897:136). Thereafter Watt rated his engines on the basis of a horsepower (hp) of 33,000 foot pounds per minute.

10.1 Nominal horsepower

Watt found that the mean pressure in the cylinders of his early engines throughout the working stroke was 7 pounds per square inch (p.s.i.) and he utilized that value in estimating power. Watt also set the value for what was then considered the "proper" piston speed in his machines at 128x $\sqrt{}^3$ of the stroke in feet per minute (Seaton, 1896:21). He then determined the power of his engines from a formula giving what was described as a "realistic unit" of power, which then used to describe the engine as being of that power i.e., its "nominal" power. This was termed "nominal" horsepower (nhp) and it was considered to be a "realistic" figure for machines in those early days (Jamieson, 1897:136; Seaton, 1896:21). Thus, the term nominal horsepower allowed engineers, manufacturers, and users to better understand the size and ability of an engine.

The formula for the calculation of nominal horsepower was: area of the piston in square inches, times effective pressure in pounds per square inch (p.s.i), times the speed of the piston in feet per minute, divided by one horsepower or 33,000 foot pounds per minute. When Watt's estimates for mean pressure and piston speed were added to the calculations the formula appeared thus:

nhp = area of piston x 7 x 128 x $\sqrt{3}$stroke divided by 33,000 (Seaton, 1896:21)

Other variables such as piston speed added to the discrepancy between the nominal power and the indicated power of an engine. By the end of the nineteenth-century the variance had become so great that for all scientific purposes the term nominal horsepower was "utterly useless" and "practically obsolete" (Seaton, 1896:22). Despite this, the term remained in use as an easy reckoner of commercial value or power of an engine based on the acceptance of the notion that the larger the machine, the more powerful it was.

10.2 Indicated horse power

In an attempt to make provision for the increasing number of variables introduced in the course of the rapid development of the steam engine, various authorities such as the Royal Navy, made numerous amendments to the original formula for nominal horsepower. It also became practice to quote the pressure of steam found in the cylinder itself. This was measured using an instrument called an *indicator* that was invented around 1780 by Watt. The product using these variables was called the *indicated power*.

The indicated horsepower (ihp) of an engine became a much more realistic guide, and according to Seaton, it fell only a little short of the actual power. Rivett (n.d.), states that ihp represents the power developed internally by an engine and that it includes the power expended in overcoming internal frictional resistance

According to Rivett, indicated horsepower is determined by substituting the mean effective pressure in the cylinder (as derived from the *indicator diagram*, which is a diagrammatic representation produced by the indicator itself), and other relevant engine data. These appear in the formula below.

ihp equals PLAN/33,000 where:

P = mean effective pressure in lbs. per square inch
L = length of piston stroke in feet
A = area of cylinder in square inches
N = of working strokes per minute.
Divided by 33,000 (one horsepower in foot pounds)

Rivett states that this is virtually the same formula as that originally used to determine nominal horsepower (nhp), but uses different expressions and actual values for piston speed and effective pressure.

In his treatise entitled *Early Marine Engineering,* Guthrie (1971:7) has suggested that if nhp is used today it is taken as one-fifth ihp. He also stated that in the case of early marine engines though "there was no clear-cut equivalent" the following ratios "will not be far out."

- Marine engines until 1820— nhp = ihp.
- From 1820 until 1840—nhp x 2 = ihp
- From 1840–1860—nhp x 3.5 or 4 = ihp

The engineer J.M.W. Sothern devotes an entire section to the description of the indicator and the analysis of the resultant indicator diagrams in his work entitled *Verbal Notes and Sketches for Marine Engineer Officers* (1939:1356–1550). Here the archaeologists or conservators may find themselves well out of their depth and may find it pertinent (as was the case at *Xantho*) to seek expert assistance, such as that rendered by our steam engineers Noel Miller, Keith Watson, and Bob Burgess.

11. APPENDIX 2: WAGES AND SALARIES IN 1870

Wages and salaries current in colonial Western Australia when *Xantho* arrived there (Knight, 1872) were as follows

Colonial surgeon £400
Postmaster General £350
Chief clerk £300
Harbormaster (Fremantle) £250
Crown solicitor £250
Headmaster (Perth Boys) £200
Draughtsman £200
Surveyor (Roebourne) £200
Harbormaster (Albany) £150
Doctor £150
Cooper and warehouse keeper £130
Teacher (Perth Boys School) £100
Teacher girls school £100
Postmaster £100
Caretaker of the public gardens £70
Hospital matron £50
General house servant £16

The scale provides a tangible indication of the worth of a ton (tonne) of pearl shell, which at the time fetched around £100–150 landed in London.

The sum of £4,500 Broadhurst outlayed for the purchase and fitting out of *Xantho* is also better understood as c. 30 times that of the annual salary of a midlevel colonial government servant. In Western Australia today that annual salary would be around AUS $50,000.

References

12. BIBLIOGRAPHY

12.1 Books and articles

Abell, W., 1948/1981, *The shipwright's trade.* Conway Maritime Press, London.

Adams, R. M., *et al.*, 1980, *Survey of the steamboat Black Cloud.* Department of Anthropology and Nautical Archaeology, Texas A&M., Texas.

Anuskiewicz, R. J., 1992, Using remote-sensing as a tool for middle-range theory building in maritime and nautical archaeology. In: D.H. Keith, and T. L. Carrell (Eds.). *Underwater Archaeology Proceedings from the Society for Historical Archaeology Conference, Kingston, Jamaica.* Society for Historical Archaeology, Tucson , Arizona: 92–99.

Archibald, E.H.H., 1968, *The wooden fighting ship in the Royal Navy, AD 897–1860.* Blandford Press, London.

Arnott, T., 1999, *Management plan for the SS Clan Ranald.* Heritage South Australia, Adelaide.

Arrhenius, O., Barkman, L., & Sjostrand, E., 1973, *Conservation of old rusty iron objects: reduction of rust with hydrogen gas.* Korrosionsinstitutet, Stockholm.

Bain, M. A., 1982, *Full fathom five.* Artlook, Perth.

Baker, Bolster, Leach, & Singleterry, 1969, Examination of the corrosion and salt contamination of structural metal from the USS *Tecumseh. Naval Research Laboratory. Memorandum Report.* Washington, D.C.

Ballard, R. D., 1987, *The discovery of the Titanic.* Hodder and Stoughton, London.

Banbury, P., 1971, *The shipbuilders of the Thames and Medway.* David and Charles, Newton Abbott, England.

Barkman, L., 1977, Preservation of large iron objects: Treatment of waterlogged finds, and conservation of finds. In J. N. Green (Ed.), Papers from the *First Southern Hemisphere Conference on Maritime Archaeology.* Ocean's Society, Melbourne: 119–128.

Barnaby, N., 1904, *Naval development of the century.* Chambers, London.

Barto Arnold, J. B., Fleshman, G. M., Peterson, C. E., Stewart, K. W., Watts, G. P., & Weldon, C. P., 1992, USS *Monitor*: Results from the 1987 Season. In J. Barto Arnold (Ed.), Advances in underwater archaeology. *Historical Archaeology: Journal of the Society for Historical Archaeology,* 26(4), University of Pennsylvania, Pensylvania: 47–56.

Barto Arnold, J. B., Oertling, T.J., & Hall, A.W., 1999, The Denbigh Project: initial observations on a Civil War blockade-runner and its wreck-site. *International Journal of Nautical Archaeology,* 28(2), 126–144.

Bass, G. F., (Ed.), 1972, *A History of seafaring based on underwater archaeology.* Thames and Hudson, London.

Bass, G. F., 1983, A plea for historical particularism in nautical archaeology. In R. A. Gould (Ed.), *Shipwreck Anthropology.* University of New Mexico Press, Albuquerque: 91–104.

Bass, G., (Ed.), 1988, *Ships and shipwrecks of the Americas: A history based on underwater archaeology.* Thames and Hudson, London.

Bearss, E.G., 1963, Underwater archaeology at the *Cairo.* In J. D. Holmquist, & A.H. Wheeler (Eds.), *Diving into the past: Theories, techniques and applications of underwater archaeology.* Minnesota Historical Society and the Council of Underwater Archaeology, Minnesota: 12–18.

Beegle, C. J., Macleod., I. D., & North, N. A., 1983, *Preliminary survey of Xantho and installation of cathodic protection system.* Report, Department of Materials Conservation and Restoration, WA Maritime Museum.

Binford, L. R., 1972, *An archaeological perspective.* Seminar Press, New York.

Binford, L. R., 1989, *Debating archaeology.* Academic Press, New York.

Bintliff, J., 1992, Comments on Thomas, J., & Tilley, C., TAG and post modernism: A reply to John Bintliff. *Antiquity 66:* 110-114.

Blainey, G., 1966/1988: *The tyranny of distance. How distance shaped Australia's history*. Sun Books, Melbourne.

Blawatsky, V. D., 1972, Submerged sectors of towns on the Black Sea Coast. In UNESCO, 1972, *Underwater archaeology: A nascent discipline*. UNESCO, Paris: 115–121.

Bourne, C.E., (Ed.), 1858, *A treatise on the steam engine in its application to mines, mills, steam navigation, and railways. by the Artizan Club*. Longman, London.

Bramwell, M., & Palmer, J., (Eds.), 1979, *The international book of wood*. Mitchell Beazley, London.

Bright, L. S., 1985, Wrecked blockade runners of the Lower Cape Fear: Site environment and physical condition. In P. F. Johnston (Ed.), *Proceedings of the Sixteenth Conference on Underwater Archaeology*. Special Publication Series, No 4. The Society for Historical Archaeology, Tucson, Arizona.

Broeze, F.J.A., 1982, Western Australia until 1869: The maritime perspective. *Early Days: Journal of the Royal Western Australian Historical Society*, Vol. VIII, RWAHS, Perth.

Broeze, F.J.A., 1998, *Island nation. A history of Australians and the sea*. Allen and Unwyn, St. Leonards, NSW.

Brouwer, N., 1985. *International register of historic ships*. Anthony Nelson, Shropshire.

Burgess, R., & McCarthy, M., 1994, SS *Xantho,* The engine model. *Model Engineer*, 172 (3970) Argus Specialist Publications, Hemel Hempstead, England: 676–79.

Burgh, N. P., 1869, *Modern marine engineering illustrated*. E. & F. N. Spon, London.

Burgh, N. P., 1873, *A practical treatise on boilers and boiler-making*. E. & F. N. Spon, London.

Carpenter, J., 1987, The uses of soil stabilizing gel media in the conservation of large and small shipwreck artifacts. *International Journal of Nautical Archaeology,* 16 (2)*:* 95-107.

Carpenter, J., 1990, A review of physical methods used to remove concretions from artefacts, mainly iron, including some recent ideas. *Bulletin of the Australian Institute for Maritime Archaeology*, 14(1), 25–42.

Carpenter, J., & Richards, V., 1994, *Conservation in Galle*. Report, Department of Conservation, WA Maritime Museum, Fremantle.

Claasen, C., 1983, Answering our questions with experiments. In R. A. Gould (Ed.), *Shipwreck anthropology*. University of New Mexico Press, Albuquerque:189–206.

Clarke, D. L., 1968, *Analytical archaeology*. Methuen, London.

Colledge, J. J., 1969, *Ships of the Royal Navy: An historical index*. David & Charles, Newton Abbott.

Commissioner of Patents, 1862, *Patents for inventions*. Eyre and Spottswoode, London.

Corlett, E.C.B., 1970, *The iron ship: the history and significance of Brunel's Great Britain*. Moonraker Press, London.

Corlett, E.C.B., 1981, Iron, steel and steam: Review paper. In D. Howse (Ed.) *Five hundred years of nautical science 1400–1900*. National Maritime Museum, London: 278–291.

Corlett, E.C.B., 1993, The screw propeller and merchant shipping: 1840–1865. In R. Gardiner, & B. Greenhill (Eds.), *The advent of steam: The merchant steamship before 1900*. Conway Maritime Press, London: 83–105.

Courbin, P., 1988, *What is archaeology: An essay on the nature of archaeological research*. University of Chicago Press, Chicago.

Cumming, D., Garratt, D., McCarthy, M., & Wolfe, A., 1995, *Port related structures on the coast of Western Australia*. Report, Department of Maritime Archaeology, WA Maritime Museum, Fremantle.

Davis, R. H., 1955, *Deep diving and submarine operations. A manual for deep sea divers and compressed air workers*. Siebe Gorman, London.

De Kerchove, R., 1948/1961, *International maritime dictionary*. (2nd ed.). Van Nostrand Reinhold, New Jersey,

Deetz, J., 1988, American historical archaeology: Methods and results. *Science*, 239: 362–367.

Delgado, J.P., 1983, Underwater archaeological investigations of gold rush era steamships on the Californian coastline. In C. R. Cummings (Ed.), 1983, *Underwater archaeology: The Proceedings of the Fourteenth Conference on Underwater Archaeology, Denver*. Fathom Eight, San Marino, CA.

Delgado, J. P., 1992, Recovering the past of USS *Arizona*: Symbolism, myth, and reality. In J. Barto Arnold (Ed.), Advances in underwater archaeology, *Historical Archaeology: Journal of the Society for Historical Archaeology*, Vol. 26: 4, 69–80.

Denny and Brothers, Ltd. 1932, *Denny Dumbarton, 1844–1932*. (2nd ed.) Burrow, Cheltenham.

Dickson, R., 1993, *Steam whistles on the Swan*. Report, Department of Maritime Archaeology, WA Maritime Museum, Fremantle.

Dingle, T., 1984. *The Victorians and settling*. Fairfax, Melbourne.

Drake-Brockman, H., 1969, Broadhurst, Charles, Edward (1826-1905). In D. Pike (Ed.), *Australian dictionary of biography, Vol. 3.* Melbourne University Press, Melbourne: 233–234.

Edington, R. S., Friend, J. H., Pineau, R., *et al.*, 1978, Some past and present shipwreck recovery projects. In *The Monitor: Its Meaning and Future,* The National Trust for Historic Preservation in the United States, The Preservation Press, Washington, D.C. 47–61.

Edington, R. S., 1978, The USS *Tecumseh.* In *The Monitor: Its Meaning and Future,* The National Trust for Historic Preservation in the United States, The Preservation Press, Washington, D.C.

Enault, P., 1988, *Alabama-Kearsage:* Duel devant Cherbourg. *La Chasse Mareé,* 3, 44–57.

Envig, O. T., 1984, Offwatch: sole surviving steam/sail ship being restored. *Marine PropulsionInternational,* February, 36–38.

Evans, U. R., 1973, *An introduction to metallic corrosion.* Edward Arnold, London.

Fincham, J., 1851/1979, *A History of naval architecture.* Scholar Press, London.

Flemming, N. C., 1988, Book review. *International Journal of Nautical Archaeology,* 17(2), 198–200.

Florian, M.L.E., 1987, The underwater environment. In C. Pearson (Ed.) *Conservation of marine archaeological objects.* Butterworths, London: 1–20.

Friend, J. H., 1978, USS *Techumseh*: Child of the *Monitor.* In, *The Monitor: Its Meaning and Future.* The National Trust for Historic Preservation in the United States, The Preservation Press, Washington, D.C.

Garcia, R., 1996, Restoring the *Xantho* engine. *Australian Sea Heritage,* 48, 16–20.

Garcia, R., & McCarthy, M., in prep., *Threads on the SS Xantho engine.* Western Australian Maritime Museum, Fremantle.

Garratt, D., McCarthy, M., Richards, V., & Wolfe, A., 1995, *The Albany Town Jetty excavation report.* Department of Maritime Archaeology, WA Maritime Museum, Fremantle.

Gaur, A. S., Vora, K. H., Sundaresh E., Tripati, T., Gudigar, P. & Bandooker, S.N., 1998, Exploration of steam engines off Minicoy Island, Laksadsweep, India. *International Journal of Nautical Archaeology,* 27(3), 225–236.

Gibbs, M., 1994, *An archaeological, conservation, and management study of nineteenth-century shore-based whaling stations in Western Australia.* National Trust for the Australian Heritage Commission, Perth.

Gibbs. M., (in prep.), *Shipwreck abandonment behavior and the material record.* James Cook University, Townsville.

Gibbins, D., 1992, Reviewing the Underwater Archaeology Proceedings from the 1990 Society for Historical Archaeology Conference. *International Journal of Nautical Archaeology,* 21(1), 82–5.

Gilbert, K. R., & Galloway, D. F., 1978, Machine Tools. In C. Singer, *et al.*, (Eds.), *A history of technology.* Oxford, Clarendon Press.

Gould, R. A. (Ed.), 1983, *Shipwreck anthropology.* University of New Mexico Press, Albuquerque.

Gould, R. A., 1990, *Recovering the past.* University of New Mexico Press, Albuquerque.

Gould, R. A., 1991, The archaeology of the HMS *Vixen,* an early ironclad ram in Bermuda. *International Journal of Nautical Archaeology,* 20(2), 141–153.

Gould, R. A., 1995, The Bird Key Wreck: Dry Tortugas National Park, Florida.. *Bulletin of the Australian Institute for Maritime Archaeology,* 19(2), 7–16.

Grantham, J., 1859, *Iron shipbuilding with practical illustrations.* (2nd ed.) Lockwood, London.

Green, J. N., 1990, *Maritime archaeology. A technical handbook.* Academic Press, London.

Gregory, A.C., & Gregory F.T., 1884/1981 *Journals of Australian explorations:1846-1858,1884.* Hesperian Press, Perth.

Gregory, D., 1999, Monitoring the effect of sacrificial anodes on the large iron artefacts on the Duart Point wreck, 1997. *International Journal of Nautical Archaeology,* 28(2), 164–173.

Gregory, F. G., 1984, *A report of the marine archaeological survey conducted by an Ottowa conservation group, SOS Conestoga during the summer of 1983.* Ontario Heritage Branch, Ministry of Citizenship and Culture, Ottowa.

Grey, G., 1843/1983, *Journals of two expeditions of discovery in the north west of Western Australia, 1843.* Hesperian Press, Perth.

Griffiths, D., 1993, Marine engineering development in the nineteenth-century. In R. Gardiner, & B. Greenhill (Eds.), *The advent of steam: The merchant steamship before 1900.* Conway Maritime Press, London: 160–178.

Guthrie, J., 1971, *A history of marine engineering.* Hutchison, London.

Guérot, M., 1988, The engagement between the CSS *Alabama* and the USS *Kearsage*: The archaeological discovery. *Mariner's Mirror,* 74(4), *355–359.*

Guthrie, J. N., Blackall, L. L., Moriarty, D.J.W., & Gesner, P., 1994, Wrecks and marine microbiology: A case study from the *Pandora. Bulletin of the Australian Institute for Maritime Archaeology,* 18(2) , 19–24.

Hadley, R. F., 1948, Corrosion by micro-organism in aqueous and soil environments. In H. H. Uhlig (Ed.), *The Corrosion Handbook.* John Wiley, New York: 486–410.

Hall, L., Hill, L. J., Warne, I., *et al.*, 1979, *The Xantho: Western Australia's First Steamship: Report of the MAAWA expedition, May 1979.* MAAWA, Fremantle.

Hamilton, J. C., 1914, *Pioneering days in Western Victoria. A narrative of early station life.* Exchange Press, Melbourne.

Hartley, C., 1982, *Western Australia, The decline of the colonial merchant shipping industry, 1870–1890: A critical biography.* Unpublished Honours thesis, Murdoch University, Perth.

Hashimoto, M., n.d., *Sunk: The story of the Japanese submarine fleet, 1942–1945.* Cassell, London.

Hawley, D., 1989, *The raising of the shipwreck Arabia.* Arabia Steamboat Museum, Kansas City.

Henderson, G. J., 1977, Developing a colonial wrecks programme in Western Australia. In J. N. Green (Ed.), *Papers from the first Southern Hemisphere conference on maritime Archaeology.* Ocean's Society, Melbourne: 69– 72.

Henderson, G.J., & Sledge, S. D., 1984, *Sunbeam*: Solving the mystery. *Western Way, 13.*

Henderson, G., 1988, The archaeological value of iron vessels and steamships. In M. McCarthy, (Ed.), 1988, *Iron ships and steam shipwrecks. Papers from the First Australian Seminar on the Management of Iron Vessels and Steam Shipwrecks.* WA Museum, Perth: 1–12.

Henderson, G. J., & Henderson, K. J., 1988, *Unfinished voyages, 1850–1880, Vol. II.* UWA Press, Nedlands, Australia.

Heppingstone, I.D. *Whaling at Port Gregory.* Unpublished manuscript, State Archives of Western Australia, Perth.

Hewitt, G., 1988, The SS *Blackbird* (1863–1878)—Wrecked at Port Albert Victoria. In M. McCarthy, (Ed.), 1988, *Iron ships and steam shipwrecks. Papers from the First Australian Seminar on the Management of Iron Vessels and Steam Shipwrecks.* WA Museum, Perth: 145–152.

Hillman, B., 1990, *The Hillman Diaries: 1877–1884.* Private Publication, Perth.

Hodder, I., 1991, Interpretive archaeology and its role. *American Antiquity,* 56(1), 7–18.

Holly, D. C., 1996, Recovery of the crosshead engine of the steamboat *Columbus. The American Neptune: A quarterly journal of maritime history and arts.* 56(1), 163-166.

Horton, D. (Ed.), 1994, *The encyclopaedia of Aboriginal Australia: Aboriginal and Torres Strait Islander, history, society, and culture.* Australian Institute for Aboriginal and Torres Strait Islander Studies, Aboriginal Studies Press, Melbourne.

Hutton, W. S., 1890, *The practical engineers handbook.* (3rd. ed.) Crosby Lockwood, London.

Hutton, W. S., 1903, *Steam-boiler construction. A practical handbook for engineers, boiler-makers, & steam-users.* Crosby Lockwood, London.

Jacobs M., Sept. 1979, From the bottom of Lake Superior. . . and off to Silver Hill. *The Smithsonian Torch.*

Jamieson, A., 1897, *A text-book on steam and steam engines: Specially arranged for the use of science and art, city and guilds of London Institute and other engineering students.* (12th ed.) Charles Griffin, London.

Jarvis, A., 1993, Alfed Holt and the compound engine. In R. Gardiner, & B. Greenhill (Eds.), 1993, *The advent of steam: The merchant steamship before 1900.* Conway Maritime Press, London: 156–159.

John Penn Project Team, 1984, *John Penn project report.* Underwater Archaeological Research Group of NSW, Sydney.

Johnston, P. F., (Ed.), 1985, *Proceedings of the Sixteenth Conference on Underwater Archaeology, Boston.* Special Publication Series, No 4., The Society for Historical Archaeology, Tucson, Arizona..

Johnston, P. F., & Robinson, D. S., 1993, The wreck of the 1848 propeller *Indiana:* Interim report *International Journal of Nautical Archaeology,* 22(3), 219–235.

Johnston, P. F., 1995, Downbound: The history of the Early Great Lakes Propeller Indiana, *American Neptune.* Fall, 1995, 323–355.

Kenderdine, S., & Jeffery, B., 1992 (Eds.)., *Muddy waters: Proceedings of the First Conference on the Submerged and Terrestrial Archaeology of Historic Shipping on the River Murray, Echuca.* State Heritiage Branch, Dept. of Environment and Natural Resources, Adelaide.

Kenderdine, S., 1995, *Shipwrecks 1656–1942: A guide to historic wrecksites of Perth.* WA Maritime Museum, Fremantle.

Kentish, P., 1995, *Stablisation of Santiago, Report No: Met, 291 Part 1.* University of South Australia, Gartrell School of Mining Metallurgy and Applied Geology, Adelaide.

Kimberly, W. B., 1897, *History of Western Australia: A narrative of her past, together with biographies of her leading men.* Niven, Melbourne.

Kimpton, G., & McCarthy, M., 1988, The freeing of the SS *Xantho* engine using a thermal lance. In M. McCarthy, (Ed.), 1988, *Iron ships and steam shipwrecks. Papers from the First Australian Seminar on the Management of Iron Vessels and Steam Shipwrecks.* WA Museum, Perth: 73–4.

La Que, F. L., 1975, *Marine corrosion, causes and prevention.* John Wiley, NY.

Lee, S., (Ed.), 1900, *Dictionary of national biography, Vol LXI.* Smith Elder, London.

Lenihan, D. J., 1983, Rethinking shipwreck archaeology: A history of ideas and considerations for new directions. In R.A. Gould (Ed.) *Shipwreck anthropology.* University of New Mexico Press, Albuquerque: 37–64.

Lenihan, D. J., (Ed.), 1989, *Submerged cultural resources study: USS Arizona Memorial and Pearl Harbor national historic landmark.* Southwest Cultural Resources Center Professional Papers, No. 23, Santa Fe, New Mexico.

Lenihan, D. J., & Murphy, L.E., 1989, Archaeological record. In D. J. Lenihan, (Ed.), 1989, *Submerged cultural resources study: USS Arizona Memorial and Pearl Harbor national historic landmark.* Southwest Cultural Resources Center Professional Papers, No. 23. Santa Fe, New Mexico: 75–113.

Lenihan, D., Murphy, L., Labadie, P., Holden, T., & Livingston, J., 1994, *Shipwrecks of Isle Royale National Park: The archaeological survey.* Lake Superior Port Cities, Duluth, Minnesota.

Little, B., 1992, Text-aided archaeology. In B. Little, (Ed.), *Text Aided Archaeology.* CRC Press, Ann Arbor : 1–6.

Lloyd's Register of British Shipping, 1869/1964, *Annals of Lloyd's Register.* Wyman and Sons, London.

Lloyd's Register of British Shipping, 1884, *Annals of Lloyd's Register.* Wyman and Sons, London.

Lorimer, M., 1988, Field techniques in the study of TSS *John Penn.* In M. McCarthy, (Ed.), 1988, *Iron ships and steam shipwrecks. Papers from the First Australian Seminar on the Management of Iron Vessels and Steam Shipwrecks.* WA Museum, Perth: 176–178.

Lyon, D.J., 1974, Documentary sources for the archaeological diver. Ship plans at the National Maritime Museum. *International Journal of Nautical Archaeology,* 3(1), 3–20.

Lyon, D.J., *1975, The Denny List.* National Maritime Museum, London.

MacDonald, G., 1994, *The little boat harbour at Port Gregory.* Private publication, Port Gregory. Western Australia.

MacGregor, D. R., 1973, *Fast sailing ships 1775–1875: Their design and construction.* Nautical Publishing, Lymington.

MacLeod, I.D., 1981, Shipwrecks and applied electrochemistry. *Journal of electroanalytical chemists,* 118, 291–303.

MacLeod, I. D., North, N. A. & Beegle, C. J., 1986, The excavation, conservation, and analysis of shipwreck sites. *Preventative measures during excavation and site protection: Second ICCROM Conference on Conservation and Archaeological Excavation, Ghent,* 113–131.

MacLeod, I. D., 1986, Conservation of the iron steam ship *Xantho*. In H. Mansell (Ed.) Towards 200, Conservation in an Australian context, Perth Conference 1985. *Bulletin of the Institute for the Conservation of Cultural Material,* 12(3,4), 67–94.

MacLeod, I. D. & Pitrun, M., 1986, The effects of microstucture on long-term corrosion. *Australiasian Corrosion Association Proceedings Conference 26*, Adelaide 1986.

MacLeod, I. D., 1987, Conservation of corroded iron artefacts: New methods for on-site preservation and cryogenic deconcreting. *International Journal of Nautical Archaeology,*16(1), 49–56.

MacLeod, I. D.,1989a, Application of corrosion science to the management on maritime archaeological sites. *Bulletin of the Australian Institute for Maritime Archaeology,* 13(2), 7–16.

MacLeod, I. D., 1989b, Marine corrosion on historic shipwrecks and its application to modern materials. *Corrosion Australia,* 14(3): 8–14.

Macleod, I. D., 1990, Conservation report on iron shipwrecks: SS *Xantho* (1872), *Bulletin of the Australian Institute for Maritime Archaeology,* 14(2), 1–8.

MacLeod, I. D., 1992a, conservation management of iron steamships–The SS *Xantho* (1872). *Multi-Disciplinary Engineering Transactions,* 1: 45–51.

MacLeod, I. D., 1992b, Report on the corrosion of iron shipwrecks in South Australia: with particular reference to the River Murray. In S. Kenderdine & B. Jeffery, (Eds.), *Muddy Water: Proceedings of the First Conference on the Submerged and Terrestrial Archaeology of Historic Shipping on the River Murray, Echuca, September, 1992.* State Heritage Branch, Dept. of Environment and Natural Resources, Adelaide: 1–14.

MacLeod, I. D., 1995, Technical Communication: *In situ* corrosion studies on the Duart Point wreck, 1994. *International Journal of Nautical Archaeology,* 24(1), 53–59.

MacLeod, I. D., 1998a, In-situ corrosion studies on iron and composite wrecks in South Australian waters: implications for site managers and cultural tourism. *Bulletin of the Australian Institute for Maritime Archaeology,* 22: 81–90.

MacLeod, I. D., 1998b, In-situ corrosion studies on iron shipwrecks and cannon. In W. Mourey, & L. Robbiola, (Eds.), *Proceedings of the International Conference on Metals Conservation, Draguignan-Figaniéres, France*, James and James, London: 116–121.

MacLeod, 1999, The lessons of the SS Xantho engine. *AICCM Bulletin,* 24, 10–19.

Main, T. J., & Brown, T., 1855, *The marine steam engine. Designed chiefly for the use of the officers of Her Majesty's Navy.* (3rd ed.) Longmans, London.

McCarthy, M., 1981, The underwater display case. *Bulletin of the Australian Institute for Maritime Archaeology,* 5: 42–45, AIMA, Fremantle.

McCarthy, M., 1982, A wreck inspection programme as an aid to the co-ordinated management of a large number of wrecksites. *International Journal of Nautical Archaeology,* 11(1), 47–52.

McCarthy, M., 1983, Wrecks and Recreation. In W. Jeffery & J. Amess (Eds.) *Proceedings of the Second Southern Hemsiphere Conference on Maritime Archaeology.* South Australian Department of Environment and Planning, Commonwealth Department of Home Affairs and Environment, Adelaide and Canberra: 381–390.

McCarthy, M., 1985, Treasure from the scrap heap: *Xantho* engine recovered after 113 Years on the sea bed. *Australian Sea Heritage,* 22–25.

McCarthy, M., 1986a, SS *Xantho*: A nineteenth-century lemon turned sweet. In P. F. Johnston (Ed.), *Proceedings of the Sixteenth Conference on Underwater Archaeology.* Society for Historical Archaeology Special Publication, No. 4, Boston : 54–59.

McCarthy, M., 1986b, The excavation and raising of the SS *Xantho* engine and Australia's first practical and theoretical seminar on iron and steamship archaeology. *International Journal of Nautical Archaeology,*15(2), News, 73–6.

McCarthy, M., 1986c, The protection of Australia's underwater sites. *Papers from the Conference on Preventative Measures During Excavation and Site Protection.* ICCROM Ghent:133–145.

McCarthy, M., 1986d, Conservators underwater (and archaeologists above). In H. Mansell, 1986, (Ed.) Towards 200, Papers from the Institute for the Conservation of Cultural Material Conference, Perth, 1985. *ICCM Bulletin,* 12(3, 4), 21–26.

McCarthy, M., 1987, The SS *Xantho* project: Management and conservation. In I. D. Macleod (Ed.), *Conservation of wet wood and metal, The Proceedings of the ICOM Conservation Working Groups on West Organic Archaeological Materials and Metals, Fremantle 1987,* WA Museum, Perth: 9–13.

McCarthy, M., 1988a, SS *Xantho*: The predisturbance, assessment, excavation and management of an iron steam shipwreck off the coast of Western Australia. *International Journal of Nautical Archaeology,*17(4), 339–347.

McCarthy, M. (Ed.), 1988b, Iron ships and steam shipwrecks. *Papers from the First Australian Seminar on the Management of Iron Vessels and Steam Shipwrecks.* WA Museum, Perth, 1988.

McCarthy, M., 1988c, SS *Xantho* excavation. In M. McCarthy, (Ed.), 1988, *Iron ships and steam shipwrecks. Papers from the First Australian Seminar on the*

Management of Iron Vessels and Steam Shipwrecks. WA Museum, Perth: 179–180.

McCarthy, M., 1989,The Excavation Continues... in the laboratory In J. Bickersteth, (Ed.), *Bulletin of the Australian Institute for the Conservation of Cultural Material,* 21–28.

McCarthy, M., 1990, *Charles Edward Broadhurst, 1826–1905, A Remarkable nineteenth–century Failure*. Unpublished, master's thesis, Murdoch University, Perth.

McCarthy, M., 1991, *The Flamingo Bay Voyage, including reports on the Japanese Submarine, I124, the iron barque, Ann Millicent, Indonesian divers at Cartier Island, and the inspection of a site believed to be the SS Koombana.* Report, Department of Maritime Archaeology, WA Maritime Museum, No. 45, Fremantle : 10–52.

McCarthy, M., 1994, Before Broome. *The Great Circle: Journal of the Australian Association for Maritime History*. 16(2), 76–89.

McCarthy, M., 1996, *SS Xantho: Towards a new perspective. An integrated approach to the maritime archaeology and conservation of an iron steamship wreck.* Unpublished doctoral thesis, James Cook University, Townsville.

McCarthy, M., 1998a, Australian maritime archaeology: Changes, their antecedents, and the path ahead. *Australian Archaeology*, 47, 38.

McCarthy, M., 1998b, The submarine as a class of archaeological site. *Bulletin of the Australian Institute for Maritime Archaeology*, 22: 61–70.

McCarthy, M., 1998c, Investigations at the Zuytdorp sites. In J. Green, M. Stanbury, M., F. Gaastra (Eds.) *The ANCODS Colloquium*. Australian National Centre of Excellence for Maritime Archaeology, Western Australian Maritime Museum, Fremantle.

McCarthy, M., 1998d, The collectors. *Historic Environment*, 14(1), 39–51.

McCarthy, M., et al., in prep., *Eliza Broadhurst.* Department of Maritime Archaeology, WA Maritime Museum, Fremantle.

McGann, S., (in prep.), Wilyah Miah, *Place of the pearl*. Department of Archaeology, University of Western Australia. Master's thesis.

McGrath, H. T., 1981, The eventual preservation and stabilization of the USS *Cairo. International Journal of Nautical Archaeology,* 10(2), 79–94.

McKinnon, R., 1993, *Shipwrecks of Kangaroo Island.* Department of Environment and Land Management, Adelaide.

Mercer, F. R., 1958, *The Life of Charles Harper.* Westralian Farmers Co-op Printing, Perth.

Merrett-Jones, M. & Pedley, S., 1998, Water quality and the degradation of submerged timbers at three historic wreck sites in the Solent. In, M. Corfield,

P. Hinton, T. Nixon, & M. Pollard, (Eds.), *Preserving Archaeological Remains In Situ. Proceedings of the Conference in April 96 at the Museum of London.* Museum of London Archaeology Service. and University of Bradford, Department of Archaeological Sciences, Bradford: 174–186.

Muckelroy, K., 1978, *Maritime archaeology.* Cambridge University Press, Cambridge.

Muckelroy, K., (Ed.), 1980, *Archaeology underwater: An atlas of the world's submerged sites.* McGraw-Hill, New York.

Murdock, L.D., & Daley, T., 1982, Progress report on the use of FMC, polysulfide rubber compounds for recording archaeological ship's features in a marine environment. *International Journal of Nautical Archaeology,* 11(4), 349–353.

Murphy, L., 1983, Shipwrecks as data base for human behavioral studies. In R. A. Gould, (Ed.), *Shipwreck anthropology.* University of New Mexico Press, Albuquerque: 65–90.

Murphy, L., 1987, Recent preservation efforts on USS *Arizona,* Pearl Harbor. In A. B. Albright (Ed.), *Underwater Archaeology Proceedings from the Society for Historical Archaeology Conference, Tucson.* Society for Historical Archaeology, Tucson, Arizona: 57–59.

Murphy, L., 1998, *H.L. Hunley, site assessment.* NPS, Submerged Cultural Resources Unit, Santa Fe, New Mexico.

Nairn, J., 1984, *Walter Padbury. His life and times.* North Stirling Press, Padbury.

Nairn, J., & Sue, J., 1975, *The ghost of the Alkimos.* WA Skindivers, Perth.

National Trust for Historic Preservation in the United States, 1978, *The Monitor: Its meaning and future,* Preservation Press, Washington, D. C.

North, N. A., 1976, Formation of coral concretions on marine iron. *International Journal of Nautical Archaeology,* 253–258.

North, N. A., 1982, Corrosion products on marine iron. *Studies in Conservation,* 27, 75–83.

North, N. A., 1984, The role of galvanic couples in the corrosion of materials on shipwrecks. *International Journal of Nautical Archaeology, 13.,* 133–6

North, N. A., and MacLeod, I. D.,1987, Corrosion of metals. In C. Pearson, (Ed.), *Conservation of marine archaeological objects.* Butterworths, London: 68–98.

North, N. A., 1997, *Review of excavation, transportation and conservation of Xantho engine.* Unpublished report for the Western Australian Maritime Museum's Departments of Conservation and Maritime Archaeology. SS

Xantho file, Department of Maritime Archaeology, WA Maritime Museum, Fremantle.

Nutley, D., 1992, *Shipwreck Atlas of New South Wales*. Department of Urban Affairs and Planning, Sydney.

Oddy, W. A.,1975, A review of procedures for the conservation of cast and wrought iron found on the seabed. *International Journal of Nautical Archaeology,* 4(2), 367–370.

Osbon, G. A., 1965, The Crimean War gunboats. Part. 1. *The Mariner's Mirror, The Journal of the Society of Nautical Research.* 51, 103–116.

Paasch, H., 1890/1977 *Illustrated marine encyclopaedia.* Model and Allied Publications. Argus, London.

Page, M., 1975, *Fitted for the voyage: the Adelaide Steamship Company 1875–1975,* Rigby, Adelaide.

Parsons, R., 1973/1980, *Steamships in Colonial Western Australia*, Private Publication, Magill, S.A.

Patents for Inventions, 1855, No 11,017, *Specification of John Penn, William Hartree the Younger, John Matthew the Younger. Steam engines, propulsiuon of vessels, &c. Eyre and Spottiswoode.* The Greater Patent Office, Holborn, London: 1–18.

Payne, P. L., 1974, *British entrepreneurship in the nineteenth-century: Studies in economic history.* Macmillan, London.

Pearson, C., (Ed.), 1987, *Conservation of marine archaeological objects.* Butterworths, London.

Pénnec, S., 1990, *La Conservation des métaux provenant de fouilles sous-marines, Rapport du stage effectué au laboratoire conservation du Western Australian Maritime Museum Fremantle, Septembre 1990.* Unpublished report, Department of Conservation, WA Museum: 85–164.

Playford, P, 1996, *Carpet of Silver. The wreck of the Zuytdorp.* UWA Press, Nedlands.

Potter, P. B., 1992, Middle-range theory, ceramics, and capitalism in nineteenth-century Rockbridge County, Virginia. In B. Little, (Ed.), *Text Aided Archaeology.* CRC Press, Ann Arbor: 9–21.

Preston, A., & Major, 1965, J., *Send a gunboat.* Longmans, London.

Purser, M., Oral History and historical archaeology. In B. Little (Ed.), *Text Aided Archaeology.* CRC Press, Ann Arbor: 25–28.

Renfrew, R. C., 1982, Explanation revisited. In R.C. Renfrew, M.J. Rowlands, & B.A., Segraves (Eds.), *Theory and explanation in archaeology: the Southhampton Conference*. Academic press, London: 1–23.

Renfrew, C., & Bahn, P., 1991, *Archaeology, theories, methods and practice.* Thames and Hudson, NY.

Reynard, K. D., Restoration of an iron star. *International Congress of Maritime Museums, 1979, Third Conference Proceedings, 1978.* Mystic Seaport Museum, Mystic, Connecticut: 16–27.

Reynolds, R., 1987, The Indernoona contact site: A preliminary report of an engraving in the Pilbara region of Western Australia. *Australian Archaeology*, 25, 1987, 80–87.

Richards, M. P., 1987, *Workhorses in Australian waters: A history of marine engineering in Australia.* Turton and Armstrong, Wahroonga.

Richardson, A. R., 1909, *Early Memories of the Great Nor-West and a Chapter in the History of Western Australia.* E. S. Wigg, Perth.

Rickover, H. G., 1976, *How the battleship Maine was destroyed.* Naval History Division, Department of the Navy, Washington, D. C.

Riley, J., 1988a, The Waterline Theory of Iron Ship Disintegration. In M. McCarthy, (Ed.), 1988, *Iron ships and steam shipwrecks. Papers from the First Australian Seminar on the Management of Iron Vessels and Steam Shipwrecks.* WA Museum, Perth: 191–197.

Riley, J., 1993, *SS Yongala: Site inspections, May 12, 1993.* Private Publication, Sydney.

Ringer, R., 1983, The marine excavation of the sixteenth-century Basque whaling vessel *San Juan* in Red Bay, Labrador: A summary of the 1982 field season. In C. R. Cummings, (Ed.), *Underwater archaeology, The Proceedings of the Fourteenth Conference on Underwater Archaeology.* Fathom Eight, Special Publication No.7, San Marino, California: 119–121.

Rivett, N., n.d., *The naval steam reciprocating engine.* The Naval Historical Society of Australia, Sydney.

Robb, A. M., 1978, Shipbuilding. In C. Singer, E. J. Holmyard, A. R. Hall, & T. L. Williams (Eds.), *A history of technology, Vol III: From the Renaissance to the Industrial Revolution c.1500–c.1750.* Oxford, Clarendon Press: 350–390.

Rodgers, B. A., 1989, Technical communication: The case for biologically induced corrosion at the Yorktown shipwreck archaeological site. *International Journal of Nautical Archaeology*, 18(4), 335–340.

Roff, W.J., 1993, Early steamships in eastern waters. In R. Gardiner & B. Greenhill, (Eds.), *The advent of steam: The merchant steamship before 1900.* Conway Maritime Press, London: 28–43.

Royal Australian Navy (RAN), Department of Defence, 1994, *Australian national tide tables.* Australian Hydrographic Publication 11, Australian Government Publishing Service, Canberra.

Ruoff, U., 1972, Palafittes and underwater archaeology. In *Underwater archaeology: A nascent discipline.* UNESCO, Paris: 123–133.

Schiffer, M. B., 1976, *Behavioral Archaeology.* Academic Press, New York.

Seaton, A. E., 1896, *Manual of marine engineering.* Chas. Griffin & Co., London.

Seaton, A. E., 1911, *Manual of marine engineering.* Chas. Griffin & Co., London.

Sennett, R., 1885, *The marine steam engine.* Longmans, Green and Co., London.

Sennett, R., & Oram, J., 1918, *The marine steam engine.* Longmans Green & Co., London.

Sequeira, C.A.C., & Tiller, A. K., 1988, *Microbial corrosion: 1.* Elsevier Applied Science, London.

Simmer, G. L. (n.d.), "Arabella" a woolship of the "nineties". *Journal of the Honourable Company of Master Mariners,* 372–4.

Simmons, J. J., 1988, Steamboats on inland waterways: Prime movers of manifest destiny. In G. Bass, (Ed.) *Ships and shipwrecks of the Americas: A history based on underwater archaeology.* Thames and Hudson, London.

Sledge, S. D., 1978, *Wreck inspection north coast.* Report, Department of Maritime Archaeology, Western Australian Maritime Museum, Fremantle.

Sledge, S., 1979, *Wreck inspection report, SS Xantho,* Unpublished wreck inspection report. Department of Maritime Archaeology, Western Australian Maritime Museum, Fremantle.

Smith, E. C., 1937, *A short history of marine engineering.* Bablock and Wilcox, Cambridge.

Smith, T., 1999, *Project AEII: Investigation of the HMA AEII submarine wreck site. NSW Heritage Office,* Sydney.

Society for Industrial Archaeology, 1979, *Indiana* salvage: Oldest American marine engine recovered. *Society for Industrial Archaeology Newsletter.* 8(5).

Sothern, J.W.M., 1923, 1939 *'Verbal' notes and sketches for marine engineer officers: A manual of marine enginering practice.* Crosby Lockwood and Son, London.

Southwell, C. R., Bultman, J. D., & Alexander, A., July 1976, Corrosion of metals in tropical environments: Final report of 16 year exposures. *Materials Performance,* 9–26.

Souza, D., 1998, *The persistence of sail in the age of steam. Underwater archaeological evidence from the Dry Tortugas.* Plenum Press, New York.

Spencer, M., 1999, The Unsung Heroine, *Sport Diving,* No. 74, June/July, 76-81.

Stanbury, M., 1993, *Historic sites of the Easter Group, Houtman Abrolhos, WA.* Report prepared for the Abrolhos Islands consultative committee. Report Department of Maritime Archaeology, Western Australian Maritime Museum, No. 68.

Stanbury, M., 1994, Mother-of-pearl cultivation: an early twentieth-century experiment in the Montebello Islands, Western Australia. *The Great Circle: Journal of the Australian Association for Maritime History.* 16(2), 90–120.

Stevens, W., 1982, Underwater stratigraphy at Red Bay: A methodological approach to explain past events. In D. H. Keith (Ed.) *Underwater Archaeology: The Proceedings of the Thirteenth Conference on Underwater Archaeology.* Fathom Eight, Special Publication No. 5, San Marino, California:17–18.

Stewart I. G., 1992, *Liberty Ships in peacetime and their contribution to world shipping history.* Ian Stewart Marine Publications, Rockingham Beach, Perth.

Strachan, S., 1999, *City of Launceston Management Plan.* Department of Infrastructure, Melbourne.

Streeter, E. W., 1886, *Pearls and pearling life.* Bell and Sons, London.

Suckling, A.J., (n.d.) *The history of the Northhampton District.* Unpublished Teacher's Higher Certificate thesis, State Archives of Western Australia.

Thearle, S.J.P., 1886, *The modern practice of shipbuilding in iron and steel.* Collins, Glasgow.

Thomas, T., & Tilly, C., 1992, "TAG and post-modernism: a reply to John Bintliff. *Antiquity,* 66,106–110.

Totty, D., 1979, *Wrecks of W.A's central mainland coast-Jurien Bay—-Port Gregory.* unpublished report, Department of Maritime Archaeology, Western Australian Maritime Museum, Fremantle.

Totty, D., 1982, *SS Xantho, Unpublished Report.* Department of Maritime Archaeology, Western Australian Maritime Museum, Fremantle.

Totty, D., 1986, *Wrecks of Western Australia's Central Mainland Coast.* Private Publication, Perth.

Trenaman, H. R., 1934, Port Gregory and Lynton. *Journal and Proceedings of The Western Australian Historical Society,* Vol. II, Part XVI, RWAHS, Perth: 1–5.

Trigger, B., G., 1989, *A history of archaeological thought.* Cambridge University Press, Cambridge.

Turgoose, S., 1989, Corrosion and structure: Modelling the preservation mechanisms, *Evidence Preserved in Corrosion Products: New Fields in Artifact Studies,* United Kingdom Institute for Conservation, Occasional Papers No. 8.

Uhlig, H. H., (Ed.), 1948, *The corrosion handbook*(9th ed.), John Wiley & Sons, New York.

USSR Ministry of Defence, 1974, *Atlas of the Oceans-Pacific Ocean.* USSR Naval Branch.

Van Doorninck, F., 1972, Byzantium, mistress of the sea. In G. F., Bass (Ed.), *A history of seafaing based on underwater archaeology.* Thames and Hudson, London: 330–641.

VandenHazel, B. J.,1987, The wreck of the side paddle wheel steamer *John Fraser,* Lake Nipissing, Ontario, Canada, *International Journal of Nautical Archaeology,*16(4), 331–341.

Veth, P., & McCarthy, M., 1999, Types of explanation in maritime archaeology: The case of the SS *Xantho. AA Australian Archaeology,* 47,11–14.

Walker, F. M., 1990, Precision construction: Iron's contribution to modern shipbuilding. *Metals in shipbuilding. Papers presented at the March 1990 Conference of the Historical Metallurgy Society,* Historical Metallurgy Society, London.

Wall, R. D., 1979, Commentary on Reynard. *International Congress of Maritime Museums, 1979, Third Conference Proceedings, 1978.* Mystic Seaport Museum, Mystic, Connecticut: 31–35.

Ward, I.A.K., Larcombe, P. & P. Veth, 1999, A new process-based model for wreck site formation. *Journal of Archaeological Science,* 26(5): 561–570.

Warsop, J. C., & Tomlin, R. J., 1990, *HMS Warrior 1860- A study of machinery installation, operation, and performance.* Institute of Marine Engineers, London.

Watson, P. J., 1983, Method and theory in shipwreck archaeology. In R. A. Gould (Ed.), *Shipwreck Anthropology,* University of New Mexico Press, Albuquerque: 23–36.

Watts, G. P., 1972, Bermuda and the American Civil War: a reconnaissance investigation of archival and submerged cultural resources. *International Journal of Nautical Archaeology,* 17(2), 159–171.

Watts, G. P., & Bright, L. S., 1973, Progress in underwater archaeology in North Carolina 1962–72, *International Journal of Nautical Archaeology,*2(1), 131–6.

Watts, G. P., 1975, The Location and identification of the Ironclad USS *Monitor. International Journal of Nautical Archaeology,*4(2), 301–329.

Watts, G. P., 1987, A Decade of research: Investigation of the USS *Monitor.* In A. B. Albright (Ed.) *Underwater Archaeology Proceedings from the Society for Historical Archaeology Conference, Savannah.* The Society for Historical Archaeology, Tucson, Arizona: 128–139.

Watts, G. P., 1988 Bermuda and the American Civil War. *International Journal of Nautical Archaeology*, 17(2), 160–171.

Watts, G.P., 1989, Runners of the Union. *Archaeology*, 32–39.

Weldon Athole c.1960. *The life and work of C.E. Broadhurst.* Unpublished Typescript, PR 4257, Battye Library, Perth.

Wise, S. R., 1985, Confederate trade and blockade running during the American Civil War. In P. F., Johnston (Ed.), *Proceedings of the Sixteenth Conference on Underwater Archaeology.* Special Publication Series, No 4. The Society for Historical Archaeology, Tucson, Arizona: 120–133.

Withnell-Taylor, N. E., 1987, *A saga of the Northwest. Yeera-Muk-a-Doo. The first settlement of Northwest Australia told through the Withnell and Hancock families 1861–1890.* Hesperian Press, Perth.

Yeo, J., 1894, *Steam and the marine steam engine.* Macmillan & Co., London.

12.2 Official documents

Correspondence relating to the loss of the steamship Xantho, and evidence at the Court of Inquiry. CSO, 727/233–275, State Archive of Western Australia (SAWA).

Glasgow Post Office Directory, 1872, Glasgow, Scotland.

Journals and Occurance Books, R.J. Sholl, Acc 194, QB SHO, SAWA.

Knight, W., Comp., 1870, *Census. Census of the Colony of Western Australia, taken on 31, March 1870 and also the general statistics of the Colony for the ten years ending 31/12/1869.* Perth, Government Print.

Liverpool Underwriters' Register, 1872, Liverpool, England.

Votes & Proceedings of the Legislative Council (V&P), *Correspondence relative to the State of Affairs on the Northwest Coast and the treatment of Malay and other labourers employed in the Pearl fishery, Votes and Proceedings of the Legislative Council, 1874–1875.*

Xantho, Certificate of British Registry. Steamers, 4/1848, Port of Anstruther, November, 1, 1848, PRO: BT 107/453. 262.

Xantho, Certificate of British Registry. May 21, 1860, Port of Leith, PRO, BT 108/67.262.

Xantho, Certificate of Registry, Glasgow, 61/1871, Port of Glasgow, PRO BT 108/250. 262.

12.3 Maps and Charts

Archdeacon, W. E., Commander, Royal Navy, 1879, *Australia West Coast, OD 97, Shoal Point 28° South to Flinder's Bay,* Admiralty Chart No. 2. Western Australian Coast.

12.4 Company Documents

Broadhurst, C. E., and Broadhurst, F. C., *Deed of Dissolution of Partnership,* Broadhurst Family Papers, SAWA.

Camden Harbour Pastoral Association, 1864, *Description of Camden Harbour, its climate and adjoining country, with map*, Fergusson and Moore, Melbourne: University of Melbourne Company Archives.

Moor Collection, *SS Xantho specifications and builders' contract.* Hertsfordshire, England. Copy on file at Department of Maritime Archaeology, WA Maritime Museum, Fremantle..

12.5 Letters

Broadhurst to Colonial Secretary. 1/7/1872, CSR 727/76, SAWA.

Broadhurst to Sholl, 14/4/1871, CSR 697/110, SAWA.

Broadhurst to Colonial Secretary 25/10/1871, CSR 1885/1182, SAWA.

Broadhurst to Colonial Secretary. 21/4/1876, CSR 488, Vol 20, SAWA.

Gratte, S., to P. Playford, 9/1/1997. Copy on SS *Xantho* file, 9/79, WA Maritime Museum, Fremantle.

Gull to Barker, March 1868, Barker and Gull letters, Acc No. 2423a, SAWA.

Government Resident's Office to Colonial Secretary, 31/12/1873, CSR 727/254.

North, N. A., To Beegle, C.J., 1984. Maritime Archaeology Dept., WA Maritime Museum file, *SS Xantho*, 9/79.

Tomlin to McCarthy, 28/12/1983. Copy on SS *Xantho* file, 9/79, WA Maritime Museum, Fremantle.

Tomlin to McCarthy, 29/9/1983. Copy on SS *Xantho* file, 9/79, WA Maritime Museum, Fremantle.

Roone to McCarthy, 18/5/1984. Copy on SS *Xantho* file, 9/79, WA Maritime Museum, Fremantle.

Willings, T., to McCarthy, 9/10/1991. Copy on SS *Xantho* file, 9/79, WA Maritime Museum, Fremantle.

Worsley, J., to McCarthy, 1983, Site conditions at the *Xantho.* Copy on SS *Xantho* file, 9/79, WA Maritime Museum, Fremantle.

12.6 Newspapers and Periodicals

*Geelong Advertiser, 1865,*Geelong, Victoria.

Glasgow Herald, 1865, Glasgow, Scotland.

*Herald, 1870–3,*Perth, Western Australia.

*Inquirer and Commercial News, 1864–1880,*Perth, Western Australia.

Perth Gazette and WA Times, 1865–1876, Perth, Western Australia.

The Engineer, Shipbuilding and Marine Engineering on the Thames in the Victorian Era, January 3, 1862; February 11, 1898; June 3, 1898, June 17, 1898).London.

UEC Newsletter, The Underwater Explorers' Club, Fremantle

West. Australian, 1886–1890, 1988,1989

12.7 Daybooks and minute books

Maritime Archaeology Advisory Committee Minute Book. Maritime Archaeology Department, Western Australian Maritime Museum.

Xantho Engine Deconcretion Book, 1985–1995. Department of Maritime Archaeology, Western Australian Maritime Museum.

Xantho Expedition Daybook, 1983–1994. Department of Maritime Archaeology, Western Australian Maritime Museum.

12.8 Personal papers

Diaries of R.J. Sholl & T.C. Sholl. RJS & TCS, QB Sho, State Library of Western Australia.

Index

Printed in the United States
96824LV00001B/78/A